T0091955

The World Before Us

'Our knowledge of where humanity came from has been revolutionized in the last ten years, and Tom Higham has been in the front lines for many of the biggest breakthroughs. *The World Before Us* is a fascinating and entertaining account, which tells us not only of how we began, but also where we might be going. If you read one book on human origins, this should be it' Ian Morris, author of *Why the West Rules – For Now*

'The application of new genetic science to pre-history is analogous to how the telescope transformed astronomy. Tom Higham, one of the world's leading scientists in the field, brings us to the frontier of recent discoveries with a book that is both gripping and fun' Paul Collier, author of *The Bottom Billion*

'Opening up entirely new perspectives on the early history of humanity, this exciting book shows that we now have a revolutionary new tool for reconstructing the human past: DNA from minute pieces of tooth and bone, and even from the dirt on the floor of caves. Everyone studying history should be taking a course in DNA' David Abulafia, author of *The Boundless Sea*

'The who, what, where, when and how of human evolution, from one of the world's experts on the dating of prehistoric fossils. Tom Higham blends evidence from archaeology, palaeontology and genetics to reveal a rich family album of our closest relatives, a cast of characters including cave-dwelling Neanderthals, mountain-adapted wanderers and island-living Hobbits, which thrived before *Homo sapiens* took over the world' Steve Brusatte, author of *The Rise and Fall of the Dinosaurs*

'Tom Higham has been at the pulsating centre of the close collaboration between archaeologists and geneticists that in the last few years discovered our previously unknown cousins – the Denisovans – and revealed the lost world in which they, Neanderthals and modern humans interacted and interbred. His thrilling book gives us a court-side view of this scientific revolution' David Reich, author of *Who We Are and How We Got Here*

The World Before Us

The New Science Behind Our Human Origins

TOM HIGHAM

Yale UNIVERSITY PRESS

New Haven & London

First published by Penguin Books Limited in the
United Kingdom and by Yale University Press
in the United States in 2021.

Yale University Press books may be
purchased in quantity for educational, business,
or promotional use. For information, please
e-mail sales.press@yale.edu (U.S. office) or
sales@yaleup.co.uk (U.K. office).

Set in 12/14.75 pt Bembo Book MT Std.
Typeset by Jouve (UK), Milton Keynes.
Printed in the United States of America.

Library of Congress Control Number: 2021936459
ISBN 978-0-300-25922-3 (hardcover : alk. paper)

This paper meets the requirements of ANSI/NISO
Z39.48-1992 (Permanence of Paper).

10 9 8 7 6 5 4 3 2 1

For Joe, Miriam, Angelo and Elektra

Contents

Contents

List of Illustrations

Plate section

A Note on Time and Dates

Dates are often mentioned in this book, so a quick word on general terminology might be useful.

The period popularly known as the Stone Age dates from approximately 3.3 million years ago until around 5,000 years ago, and thus covers more than 99 per cent of human technological prehistory. It is divided into three parts: the Old Stone Age, or Palaeolithic, the Middle Stone Age, or Mesolithic, and the New Stone Age or Neolithic. The Old Stone Age is further divided in the Lower, Middle, Upper and Late Palaeolithic periods.

The dating of these periods differs in different parts of the world.

The Palaeolithic begins with the earliest stone tools, which currently are dated at around 3.3 million years ago in Africa, at the Kenyan site of Lomekwi. The Middle Palaeolithic begins 300–350,000 years ago and ends broadly between 40,000 and 50,000 years ago depending on which region of the world is being considered. It embraces the period of the Neanderthals as well as the development of tools made using a new approach called Levallois, in which stone tools became markedly more efficient in terms of the size of their cutting edge. The Upper Palaeolithic follows and has often been exclusively linked with our own species, although this is not likely to be strictly true any more.

This book focuses in the main on the Middle and Upper Palaeolithic, and importantly, on the nature of the transition between the two.

The Mesolithic period starts around 15,000 years ago, at the end of a long period of extreme glacial cold known as the Ice Age. As the climate of the world warmed, a range of new habitats opened up and humans moved into these environments armed with distinct new stone-tool technologies. From these new environments they exploited a wider spectrum of food resources such as small mammals and marine foods.

The Neolithic, which begins around 10,000 BC, is characterized by the development of agriculture, which spread from several major centres depending on the crops and animals that were domesticated. The Neolithic is often associated with polished stone tools, a sedentary lifestyle, rather than hunting and gathering, and ceramics or pottery.

1. Introduction

Monday 22 June 2015 at 9:10 am. One of the great moments of my life. I was in one of the laboratories at the Research Lab for Archaeology at Oxford University, where I have worked for the last twenty years. With one of my students, Samantha Brown, I was about to pick up for the first time the tiny bone of a human being who lived around 120,000 years ago.

We had found it, one bone amongst tens of thousands of other fragments, using a brilliant new scientific approach called ZooMS, which is an acronym for zooarchaeology by mass spectrometry. Sam's persistence over weeks, in taking minute samples from over 1,500 tiny bone fragments for analysis from the site of Denisova Cave in Siberia, had paid off.

The bone was tiny, only 2.4cm long, but, as we later discovered, very, very special. It is, so far, the only existing bodily remnant of a person who was a genetic hybrid: the offspring of two different groups of humans. This young woman's mother was a Neanderthal and her father a Denisovan, a distinct group of humans that was discovered only in 2010 by geneticists at the Max Planck Institute in Germany when analysing material from the Denisova site. Think of them as our distant cousins, and the closer cousins of Neanderthals, who lived mainly in Europe and the Levant between 250,000 and 40,000 years ago.

This tiny bone represents the first time anyone has identified a first-generation (F-1) hybrid in archaeology. It has made us think about how often such events might have occurred between peoples in the deep past and question what the species designations mean when it comes to different groups of humans. How can we really say two species are different if, as it seems from this finding, they can interbreed successfully?

Finding her was incredibly lucky, but, as the saying goes, you make your own luck. It relied, like so much in the world of palaeo-anthropology, on collaborative teams of archaeologists and scientists using a range of cutting-edge scientific methods that together are opening new insights into the story of early humanity.

This book explores the Palaeolithic era (or Old Stone Age), a key phase of late human evolution from roughly 300,000 to 40,000 years ago; a period when we, *Homo sapiens*, became us. This area of research has changed dramatically over the last couple of decades, and what we know now about our own deep past is very different from what we once thought. This is a story of archaeological research, often in difficult terrain, coupled with brilliant new laboratory methods that together help us to answer those most fundamental of questions – where did we come from and how did we become human? It is also a story about lucky finds, often made by non-specialists and members of the public: collectors, miners, hunters, fishers, people who noticed something unusual – a bone, a piece of jaw or a partial skull fragment – and passed it on to an expert. Some of the fossils concerned are now amongst the most important that we have in palaeoanthropology.

Unlike more recent archaeological periods, in which we are for-tunate to have evidence for towns, cemeteries, houses and the detritus of household rubbish – pots, animal bones, metal and so on – evidence from the Palaeolithic is often fragmentary and poorly preserved; tiny pieces in a jigsaw puzzle that we can never quite complete. The Denisovans themselves are a case in point. In 2020 only six biological samples from this population existed: three teeth and three bone fragments (as well as the hybrid bone, of course, which I suppose counts as half a Denisovan). There are no complete skulls or skeletons. Despite this, there is a significant amount that we already know and can glean from the evidence recovered thus far. Much of this comes from the field of ancient genomics, a ground-breaking technique that has allowed us to delve into the molecular evidence for Denisovans and explore aspects of their population his-tory that have implications for us as well as Denisovans. Events in this period of human evolution, however, can move quickly and a single discovery can really change our interpretations and the way we think about what happened in the past. This makes such explorations very exciting. Recent discoveries have added hugely to what we can say about Denisovans and their way of life, their geographic distribution and their contribution to aspects of our modern world.

I have been working as part of the team at Denisova Cave for the last few years, responsible for dating the site and the archaeological remains in it, as well as working with others to discover more human bones like the tiny hybrid bone. Part of the story I want to tell is of the site and the astonishing archaeological and genetic discoveries that have come from it.

But the Denisovans are one strand of a much bigger story. What we know about the evolution of our genus, *Homo*, has changed dramatically over the last two decades. Research has shown unequivocally that the Earth was a primevally complicated place 50,000 years ago. To borrow from the words of Tolkien, we should think of it as a veritable 'Middle Earth' in terms of the diversity of forms of the human family that existed at the time. There were five, six, or even more, different types of human present in various parts of the world. I aim to widen the story of human evolution and explore who these different groups were, as well as asking why it is that we are the last ones left.

It is most important to know about our ultimate origins, so in Chapter 2 we will meet our earliest human ancestors, people who evolved in Africa some 250–300,000 years ago, and explore when they moved outwards and into the rest of the world. Do not for a second imagine that this story of our African origins is simply one of us evolving there and later expanding outwards. We will see that at the time of our early evolution we were almost certainly not alone in that continent; there were other human lineages living there too, and they probably overlapped with us both geographically and temporally. We will discover who these other people were and what kind of contact we might have had with them.

Following their African exodus, our ancestors encountered other kinds of humans. In Europe, the Levant, Central Asia and the Altai Mountains, lived the Neanderthals, our best-known relatives. Later, as the human story expands into the east of Eurasia and Southeast Asia (Chapters 4 and 7), we will learn of other, more recently discovered members of our human family, including the Denisovans, of course, the enigmatic Hobbits (*Homo floresiensis*), who lived only on the island of Flores in Indonesia, as well as a new human relative on the island of Luzon in the Philippines, discovered as recently as January 2019 (Chapter 12). We will also meet *Homo erectus*, a much more ancient human lineage dating back about 1.6 million years, and consider whether it might have survived later than anticipated and perhaps overlapped with our modern human ancestors when they first entered Island Southeast Asia (Chapter 14). We will follow the footsteps of our human ancestors as they moved into new environments and unfamiliar lands for the first time; to Australia and New Guinea (Chapter 13), the rainforests of South Asia and Sumatra, the temperate northern regions of Siberia and more (Chapter 10). What did these people need to do to survive in these new places? What was the effect of the climate and how different did the world look in these ancient times?

We will consider what happened when these various groups met one another in the world before us. Was there contact, and, if so, how much? Did we exchange genes? Did we share ideas and culture? Do we inherit a cultural and genetic legacy from these ancient disappeared humans? Or did we simply extirpate them on our journey to become the last humans on earth? What happened to these lost relatives (Chapter 15)?

When I started my doctorate in July 1990, I remember standing in a white coat in a chemistry laboratory in front of a bank of evacuated glass vessels and Bunsen burners, 'doing radiocarbon dating'. It was the most incredible feeling. I remember looking up at the crazy science lab in front of me and shaking my head in wonder, thinking what power there was in this science, that you could date events in human prehistory that took place 10,000, 20,000, 30,000 or more years ago. I was hooked.

I have always been fascinated by the past – my father is an archaeologist – and I am very fortunate to work now at the University of Oxford in one of the founding archaeological science facilities in the world and a hothouse for the development of new methods to understand the past.

Archaeology in the twenty-first century is getting more and more exciting, because there is so much more that we can now discover, even from tiny fragments of material. Archaeology is a seriously multidisciplinary endeavour, bridging the sciences and humanities. It is reaping the benefits of a surge in scientific developments in a range of fields over the last thirty-odd years. Gone are the days of lone explorers or small groups of archaeologists digging up material and reporting their finds to colleagues behind closed doors or in sober monographs and reports. In order to do a proper job, careful and rigorous post-excavation analysis is required. A wide variety of expertise is needed. No one person can do it, and so it is crucial to build collaborative units and work together. Archaeology really is a team game.

The scientific arm of archaeology accounts for an increasing majority of all of the publications in the wider field. Radiocarbon dating, the game-changing chronometric method that heralded the birth of archaeological science in the early 1950s, is a tool now used in over a hundred laboratories around the world. It allows us to date events up to just over 50,000 years ago. As we will see in Chapter 9, by incorporating radiocarbon measurements with a method called Bayesian statistics we can provide much more precise estimates of when something took place. In the most recent periods (<10,000 years ago) we can determine age down to the precision of single generations. We can date anything that once lived using radiocarbon, but other techniques allow us to date inorganic samples too. Single grains of quartz and feldspar minerals can be dated using methods that can translate the amounts of radioactivity that have been taken up in their crystal lattices over millennia into estimates of elapsed time. We will see that isotopes of uranium and thorium can also provide chronometric information that can help us to date teeth and bones, as well as the minute build-up of calcium carbonate over ancient rock art made by humans.

Analysis and measurement of the isotopes of carbon, nitrogen, strontium, oxygen, sulphur and more, can tell us about the types of food that people and animals consumed and the changing temperatures and climates they lived in throughout their lives. In Chapter 3 we will see how this has helped us to decipher minute details of

Neanderthals' lives and diet. We can tell when someone stopped and started eating different types of food; when they moved from one place to another to live; and when they were affected by environmental contamination and by how much.[1] We can determine when an infant stopped being weaned, based on the changes in the elemental and isotope signals in their milk teeth.* We can identify periods of stress by measuring the presence of accentuated lines in tooth enamel. At the site of Payre in the Ardèche of France, for example, a Neanderthal tooth showed a week of stress at 701 days of age that occurred in the coldest part of the winter.[2]

Ancient dental plaque can tell us about diet and aspects of the bacterial colonies that inhabited the mouths of prehistoric people, their so-called microbiome. This contains an archive of disease, infection, bacteria, viruses and the vicissitudes of daily life that can be analysed using DNA, high-resolution microscopy and a new science called 'proteomics'. CT scanners are used in hospitals to assess patient health, but they can also be used to peer into ancient bones and teeth, determining age, aspects of health and periods of stress in more ancient times. In Chapter 6 I will describe how scientists used CT scans to explore the bone density of a tiny piece of Denisovan finger bone and thereby worked out that it came from the right hand of a girl of 13.5 years of age. Geometric morphometric analysis can be

---

* The ratio of the isotopes of nitrogen, ^{14}N and ^{15}N, increases with biochemical processes as they pass up successive trophic levels; from plants to herbivores to carnivores. High trophic level animals such as carnivores have elevated isotope ratios compared with the herbivores they consume. An infant in the womb will have the same value as its mother, but after birth, as it consumes breast milk, its values will increase by 3 to 5 parts per thousand, because it is effectively a trophic level higher. As the baby is weaned its values drop back to a lower level, the same as the mother if their subsequent diet is the same. By measuring these isotopes in bone and hair, where preserved, one can estimate the date of weaning. Using other techniques and materials such as teeth, it is possible to be very precise in these estimates. Using high-resolution CT scans, researchers can work out the exact age of a child based on daily growth lines. Measuring the abundance of barium and calcium in the teeth can also reveal when the child was weaned. Elevated values reveal the initiation of milk consumption. Oxygen isotopes reveal the passing of seasons because these are temperature dependent, as we shall see in Chapter 4.

applied to skulls of both animals and people to compare differences in shape and determine subtle changes between them, plotting them in different dimensions to measure relatedness. 3D modelling enables us to visualize these shapes and flip and turn them in virtual space, as we will see in Chapter 16 when looking at the extent to which our modern skulls carry the genetic influence of Neanderthals.

We now talk about building 'life histories' from fragmentary human remains because by using scientific methods such as these we can understand so much about when, where and how people once lived.

Scientific methods can also help us to study the range of different materials that are excavated. Geochemistry can help us source the places where rocks and stones were mined before being made into stone tools. We can trace the distance people must have travelled to source such rocks, or trade them with others. We use computerized shape analysis to study the variety in different stone tools and categorize them using complex statistical packages.

The minute vestiges of cut marks made by humans in the past, processing joints of meat and carcasses, can be studied under high-resolution and scanning electron microscopes. In Chapter 9 the importance of this will be shown when we attempt to build a chronology for the Denisova site and work out when humans were present at the cave.

Using drones, satellites and LIDAR (light detection and ranging: using lasers to scan the surface of the Earth in 3D), we can look through forest cover and map ancient sites and landscapes from far above. We now talk about 'digital' or 'cyber archaeology' to describe these approaches to understanding the past. Geophysical surveying using ground-penetrating radar and magnetic imaging can even allow us to look into the earth beneath our feet for anomalies that indicate the presence of archaeological features which can then later be excavated.

Specialists in archaeological teams work on identifying animal bones, botanical remains, pollen spores, sediments, faecal biomarkers, organic residues and so on, to reconstruct human adaptation and the changing climates and environments of the past. Ancient DNA methods allow us to track genetic admixture and introgression between populations, the relatedness of people buried in cemeteries and their population history. There even appears to be a way of reconstructing a person's physical appearance based on the so-called 'epigenetic' patterns in the genes we carry, as we will explore in Chapter 7. There is almost no end to the imaginative ways in which we can apply science to the past. Discovery is now as common in the laboratory as it is through digging at the point of a trowel. In this book I hope to convey some of the excitement of the latest discoveries in the field from the perspective of the people actually doing the work, whether that is at the archaeological site or in the laboratory.

Nothing would be possible, however, without the careful and rigorous excavation work of teams like those at Denisova Cave. In archaeology, *context* is everything. Knowing the precise locations of different items that are recovered is key to placing the results from post-excavation work into their proper perspective and putting the complex, never-to-be-completed jigsaw puzzle of the past together. This is where archaeology starts – in the field in the one-time-only process of excavation. Without archaeology and excavation there would be no scientific study of the material remains from those sites. Fortunately, there are many excellent archaeological teams that work around the world in time periods ranging from the modern era all the way back to the beginning of the human lineage.

Of course, any story must start at the beginning, and in the case of all humans in the world today, our story starts in Africa.

2.   Out of Africa

The idea of an African origin of humanity dates back to Charles Darwin, who predicted that in order to find the ancestors of humans we ought to explore in those places where our closest living relatives, the great apes, now live. It was not until the 1920s, however, that palaeoanthropologists began to become aware of its fossil hominin record. In 1921 miners discovered an ancient skull at a place called Broken Hill (now called Kabwe) in Zambia. The skull

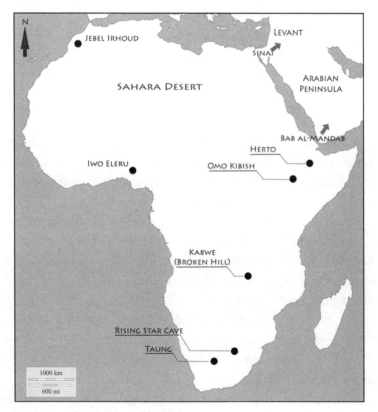

Figure 1 African sites and locations.

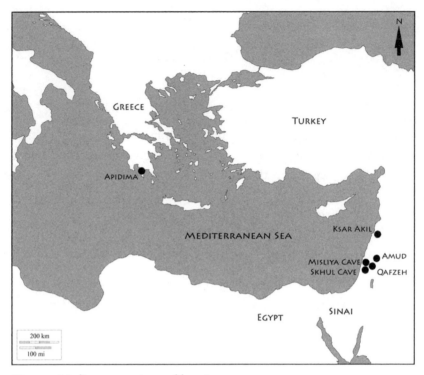

Figure 2 Mediterranean sites and locations.

was given to the British Museum in London, where it was identified as a new and ancient species called *Homo rhodesiensis*. Shortly after, in 1924, Raymond Dart found the tiny preserved skull of a 2 million-year-old *Australopithecus africanus* – the famous Taung Child – in South Africa (although several scholars refused initially to see its significance). In the years since, the fossil record of our genus, *Homo*, and other hominin relatives, has been added to considerably. Dozens of research teams, working in all parts of Africa, most notably in the eastern, southern and north-western regions, have discovered a wealth of new fossil evidence. We know now that our genus emerged some 2.5 million years ago in Africa, and that later members of *Homo*, including *Homo sapiens*, evolved subsequently in the same continent.

The primacy of Africa in our story emerged amongst the research community gradually in the 1970s and 80s, as increasingly quantitative methods were brought to bear on the study of cranial remains of *Homo sapiens* and our near relatives.[1] Researchers such as Chris Stringer of the Natural History Museum in London coined the term 'Recent African Origin' (or 'Out of Africa II') to describe a model explaining the recent origins of all people beyond that continent and the earliest appearance of modern humans in Africa. As the name implies, the movement of our ancestors out of Africa is not the first time this happened. Out of Africa I refers to the dispersal of an earlier group of humans, *Homo erectus*, into Eurasia, some 1.6 million years ago. Although below the neck they looked very similar to us, their brain size was much smaller – between 650 and 800cc (cubic centimetres) in the earliest individuals. In a remarkable and very successful radiation, they expanded all the way to Island Southeast Asia, where we find their remains by about 1.5 million years ago.

It is likely that there were other movements of people out from Africa too at different times. Neanderthals and ourselves, for example, have a common ancestor. Precisely who this was is not completely clear, but it is likely that this ancestor is African and dates from at least 530,000 years ago, based on genetic estimates.[2]

Other researchers considered alternative models more appropriate to explain the patterns in the fossil record. The multiregional model held that modern humans appeared in different regions of the world following the dispersal of *Homo erectus* into Eurasia.[3] Interbreeding, with periodic gene flow between groups was later argued to have occurred, ensuring that complete separation between these various regional populations did not take place and humans evolved broadly in parallel.

The interpretation of an African origin for all humanity, however, was spectacularly confirmed in a landmark paper in the scientific journal *Nature* in 1987. The authors* analysed 147 mitochondrial genomes from modern human groups and found that the highest genetic diversity was in Africa. The evidence supported a model in which Africa was the most likely source of the human mitochondrial gene pool.[4] As one moves further and further away from Africa, so

---

* Rebecca Cann, Mark Stoneking and Allan Wilson.

the genetic diversity one measures in different human populations declines with a high linear correlation in the region of 90 per cent. This is an example of the classic 'serial founder effect' in biology. As smaller groups bud off and colonize new places, it follows that there must be a reduction in genetic diversity, broadly along the lines of the reduction in the founding members of new colonies.[5] Archaeological excavations and the discovery of a number of human fossil remains, along with the application of new dating techniques, has confirmed that modern humans appear to date earliest in Africa.

Intriguingly, we can see support for this too from the evolutionary history of other organisms that tend to follow humans as they move, so-called human commensals. Bacteria are a good example. They also show the same disparity in genetic diversity between African (more diversity) and non-African (less diversity) forms. Take *Helicobacter pylori*, the stomach bacterium that causes ulcers and gastric cancer. When we plot genetic diversity in this bacterium, we see a similar phylogenetic tree as in the human tree, with Africa sitting basal to everywhere else.[6] This suggests that when people left Africa, they were already infected with *H. pylori*. A small budding group of humans took with them a bacterium with lower genetic variation than the larger African host population, undergoing the same serial founder effect as their human hosts who carried them along, further and further away from Africa.*

What did these early humans look like? Were they identical to us? And when, precisely, do they appear in the fossil record?

Perhaps we should first define what we mean by 'modern human'. The majority of palaeoanthropologists agree that we are characterized by a small face, the presence of a chin and a braincase that is more 'globular' in shape than those of other hominins in the fossil record. A larger cranium is also a characteristic; over the long duration of human evolution we see a steady increase in cranial capacity. In addition, there seems to be a reduction in prognathism, which refers to the protrusion of the jaw and 'snout'.

---

* The malarial parasite *Plasmodium falciparum* exhibits the same Out of Africa pattern as *H. pylori*.

Not surprisingly, identifying the precise point at which these traits become fixed in us, and early humans become 'modern' in appearance, is an exceedingly difficult task because the fossil record is incomplete and patchy. When we look at people around the world today, we see diversity and variation. Diversity in body type, skin colour and skull shape, as well as culture and language, of course. It was the same in the past; there is a high degree of morphological variation amongst the human remains that fall into the period from about 300,000 years ago.[7]

Jebel Irhoud, in Morocco, is one particularly early and important site for tracing human origins. In 1961, miners working to extract barite there found a beautifully preserved human skull. Later, in 2004, renewed excavations revealed more remains from five different individuals. The Jebel Irhoud people had large, elongated, rather than globular, braincases, with short faces retracted underneath.[8] New dates from near the human remains indicated an age of approximately 300,000 years,[9] much older than had been anticipated. Jebel Irhoud looks as though it represents some of the earliest evidence for our species on its pathway to becoming anatomically like us. For this

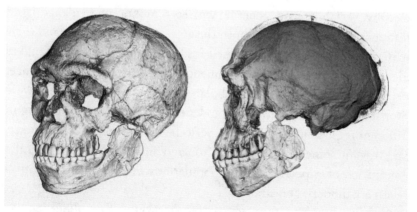

Figure 3 The first of our kind. The earliest known *Homo sapiens* fossils, from the Jebel Irhoud site. These images are composites derived from micro-CT scans of multiple original fossils. The modern-looking face falls into the range of variation one sees in living humans, but there are archaic features in the braincase (shaded in the right hand image) suggesting that the shape of the brain had not yet reached that seen in modern humans.

reason, researchers often term these people 'archaic *Homo sapiens*' rather than anatomically modern humans.

Other important fossil remains, dating somewhat later, come from Ethiopia. In 1967, at two sites in the valley of the Omo river, the palaeoanthropologist Richard Leakey found two partial skulls in the Kibish geological formation. They were later dated to 195,000 years ago.[10] They have large braincases, reduced brow ridges and wider skulls around the temple area, like us. They differ slightly from one another in terms of their shape. North-east of Omo Kibish in the Awash Valley, three more well-preserved human crania were discovered in 1997 at the site of Herto by Tim White's team. They revealed a more modern appearance, but also subtle differences and variations. They date a little later, to around 150,000 years ago.[11] They are thought by some to be intermediate between more archaic forms of *Homo* in Africa, such as the Jebel Irhoud specimens, and modern humans: 'a population that is on the verge of anatomical modernity but not yet fully modern'.[12]

These key fossils, and others like them, show that while humans in Africa are on an evolutionary trajectory towards modern people, this does not seem to have been linear or rapid; it is a trajectory characterized by variability. In general, we see a mosaic of evidence from different parts of Africa, in both the appearance of humans and their material culture, painting a diverse multiregional picture. This makes identifying precisely when our phenotypes, or outward appearance, became fixed, very difficult. Scholars have suggested that there were probably phases of isolation and independent evolution happening in different regions in Africa, with periodic interbreeding and contact, and then at some point between 100,000 and 150,000 years ago the emergence of a population or populations of *Homo sapiens* with a relatively modern appearance.

Stone-tool evidence from across Africa shows that from about 300,000 or so years ago, change was also happening culturally. At about the same time as the people from Jebel Irhoud appear, the hand axes that had dominated the stone-tool inventory over the preceding million years or more become much more rare. In their place we find a range of stone tools made by shaping flakes chipped from a large core.

(These had previously been discarded as waste.) This period is known as the Middle Stone Age, or MSA, and it begins everywhere in Africa at an almost identical time. Archaeologists can discern subtle regional differences in the stone tools produced, as humans experimented and developed new technical approaches to toolmaking. Later, as we move forward in the archaeological record, we see more complexity, a widening of materials to encompass bone and wood, the arrival of smaller blades and bladelets, the use of hafting, the presence of engraved markings found on eggshells and ochre, and ornaments made of pierced shells. Cognitive developments almost certainly underlie these changes in material culture.[13]

It has also proven difficult to identify the one group or subpopulation within Africa that was subsequently ancestral to everyone outside it. It is probably unlikely that we ever will. Once again it seems more likely that there was gene flow, population movement and mixing between groups across wide areas of Africa through time. As we shall see again and again in this book, the picture is made even more complex by the high chance that this diversity in human groups was magnified by the presence of other *types* of humans in Africa too. We, *Homo sapiens*, were not alone on that continent.

This was brought into stark contrast in October 2013 in a cave in the Cradle of Humankind World Heritage site in Gauteng, South

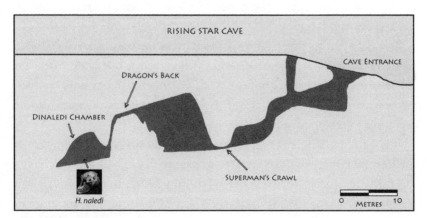

Figure 4 Plan of Rising Star Cave, with the location where the skeletal remains of *Homo naledi* were found.

Africa. Archaeologists led by Lee Berger found human remains here at one of the richest human fossil sites ever located. Cavers had originally identified the so-called Dinaledi Chamber of Rising Star Cave and alerted Berger. Access is extremely difficult. At two points between the chamber and the outside world one has to squeeze through tiny constrictions. One of these is called 'Superman's Crawl', because you can traverse it only by stretching one arm in front of you as you push through. The major challenge comes later, though, when you have to squeeze yourself through a hole that measures only 18cm across. Understandably, only the very slimmest can make it into the chamber. The team were forced to advertise for six lithe female archaeologists to do the excavation work there. These 'underground astronauts', as they were known, spent hours excavating each day. Their efforts were worthwhile: more than 1,500 human remains were recovered in November of that year, and another 1,700 bones early in 2014. Of the 206 bones found in the human body only twenty were not found in the remains of the Dinaledi Chamber.[14] The bones belong to a dozen individuals of a primitive hominin species that stood around 1.5m in height and weighed 40–55kg. Its brain size was small: between 460 and 560cc.[15] A mosaic of features is evident amongst the remains, some *Homo*-like, others similar to the much earlier Australopithecines. The crania are more similar in shape to early *Homo* remains, and for this reason the finds have been assigned to that genus and given the name *Homo naledi* (*naledi* means 'star' in the local Sotho language).

The date of the cave and its contents was initially a mystery, with many estimating an age of several million years, but later work has suggested a surprisingly recent date: between 236,000 and 335,000 years ago.[16] This means that our early *Homo sapiens* ancestors almost certainly overlapped with *Homo naledi* in Africa.

It is quite possible that there were other late surviving members of the human family in parts of Africa that are yet to be found archaeologically. In 2020 scientists reported finding DNA in modern African people that could not be sequenced to any other known human genome. This is taken as possible evidence of a 'ghost' population that must have lived in Africa tens or hundreds of thousands of years ago.[17] (We will hear more about other possible ghost populations later in the book when we consider the question of the late survival of *Homo erectus* in Southeast Asia.) The enigmatic Iwo Eleru skeleton is another interesting and difficult to explain case. Found in a cave in south-west Nigeria in 1965,[18] the skeletal remains, including well-preserved parts of the cranium, appear distinct from recent African skull shapes and sit intermediate between modern humans on the one hand and Neanderthals and *Homo erectus* on the other when compared. The estimated age is again surprisingly recent: between 11,000 and 16,000 years. The dearth of other human fossil remains in western Africa makes it difficult to place this specimen in its wider context and one hopes that renewed fieldwork in this key region will shed more light on these questions in future. In 2020, the enigmatic Kabwe skull that was found in 1921 in Zambia was directly dated. This specimen has been strongly linked with a hominin called *Homo heidelbergensis*, one of the potential common ancestors of us and Neanderthals. It produced an age of 299,000 ± 25,000 BP (Before Present), implying strongly that it overlaps with the ages of early *Homo sapiens* and *Homo naledi*[19] and therefore may not in fact be on our direct evolutionary line. This implies that there may have been more than three overlapping human groups in Africa 200–300,000 years ago.

I remember studying human evolution as a student in the 1980s. It was so simple and linear then; the story is now much more complex, and infinitely more interesting and puzzling.

Despite knowing that Africa is our ultimate origin, it is much less clear when the ancestors of people living outside Africa today left, or indeed why they left. There is much speculation over this, but equally a great deal of evidence that can be gleaned from the archaeological, genetic and palaeoenvironmental records. There are two main hypotheses for when Out of Africa II happened. The first, and more orthodox, account is that it occurred around 50–60,000 years ago.[20] Some proponents of this idea suggest that something must have happened to humans around this time, perhaps to do with cognitive changes, that conferred an advantage, enabling modern humans to expand beyond their African range.[21] The evidence for an exit at this date is based on a combination of dated archaeological sites and age estimates for mitochondrial DNA trees based around a haplogroup called L3.* This haplogroup is found widely in sub-Saharan Africa and also in all ancient non-Africans, suggesting that it may represent an ancestral group of people who left Africa.[22] Age estimates for this have coalesced around 60–70,000 years ago. Critics suggest that this assumes the demographic history of humanity has been tree-like, with little or no gene flow.[23] We do know, however, that Australia was settled at a broadly similar time, and this has acted as a minimum age for the expansion of humans from Africa, since it must have been before this date. Recent work has suggested a date perhaps as early as 59–65,000 years ago for Australian colonization[24] (we will explore more on this in Chapter 13).

The second Out of Africa II scenario proposes a much earlier exodus of modern humans, perhaps 120–130,000 years ago or earlier. There is evidence for people occupying the Near East at this time at key cave sites such as Qafzeh and Skhul in Israel. Interestingly, it seems that later, around 60–70,000 years ago, they disappeared from the region to be replaced by Neanderthals – as seen at other nearby sites like Amud. Analyses of stone-tool industries suggest potential links between the lithics of eastern Africa and those from further north, including the Levant at about 120,000 years ago,[25] implying a potential link in material culture as people moved out of Africa.

Elsewhere across Eurasia the evidence for a very early modern human presence is a little less clear and a bit more controversial. In

* Haplogroups are the different maternal groups or clades that form the branches of a tree of mitochondrial relatedness. Each maternal haplogroup can be traced to a single person who lived at some point in prehistoric time. Haplogroups can therefore be informative in exploring the female line of descent and the split times of the various haplogroups. We now know, however, that human evolution is characterized by punctuated admixture and gene-flow events between different groups. For this reason, the true story of human evolution is much more complex and the full story more effectively revealed by genome-wide analysis.

China, for example, there has been excitement about the site of Fuyan Cave in Hunan province, where forty-seven teeth have been dated to more than 85,000 years ago.[26] The teeth are dated with reference to a layer of carbonate that is argued to cover them, rather than dates on the teeth themselves. I think direct dates from these teeth are needed, so we must be cautious over whether they really are as old as claimed. We tried to date bones from the Fuyan site in my laboratory, but there was no protein remaining and so our attempt failed. Direct dating will therefore probably have to be undertaken using another method.

There are other sites in eastern Eurasia that also contain evidence for humans, but whether they are archaic forms of *Homo sapiens*, modern humans, or something else is not clear in most cases. The picture is muddied by the challenges of dating, the difficulties in extracting DNA from human remains in warmer and tropical environments and the fragmented and poor nature of many of the fossil remains. What is often lacking is the smoking gun: an unequivocally well-dated modern human from an unimpeachable archaeological context.

New archaeological evidence from the west of Eurasia, however, has provided new evidence that suggests that early *Homo sapiens* were able to move out of Africa at surprisingly early periods. Again, though, much hinges on the reliability of the dating.

In 2019 a reanalysis of material from an old archaeological site in Greece on the Mani Peninsula in the Peloponnese was published. Apidima Cave was excavated in the 1970s and a partial cranium called Apidima 1 was discovered encased in rock. It was later transported to Athens for safekeeping. Unfortunately, the curator in charge of the specimen forbade anyone to actually study it. Thus, it had lain there for more than forty years until Katerina Harvati of Tübingen University and her colleagues finally obtained permission to start working on it.[27] Using the latest CT scanning approaches, they analysed the informative rear of the cranium and created 3D models of it. They showed that the find was more similar to the skull shapes of other *Homo sapiens* than to those of Neanderthals. Dating of the specimen using uranium-series methods showed it was *at least* 210,000 years old.* The partial nature of the skull urges us to be cautious regarding the diagnosis that it is *Homo sapiens*. One hopes more evidence will be forthcoming to add confidence. What was very interesting about Apidima 1, however, was that near it lay a second cranium, Apidima 2. This one, according to the team, fitted with a Neanderthal-type morphology. It was also slightly younger, at more than 170,000 years. The evidence therefore suggests that there were two hominin groups in the region, albeit at different dates.

Another site in Israel, Misliya Cave, has also produced evidence for very early *Homo sapiens* outside Africa.[28] The site is on Mount Carmel, which is dotted with important cave sites and which has been the focus of archaeological research for decades. A jawbone with associated teeth was found at Misliya in 2002. Dating methods suggest that the specimen is between 177,000 and 194,000 years old – again, substantially older than either of the two widely cited dates for human expansion into Eurasia.

The evidence suggests, then, that there were at least two movements of *Homo sapiens* out of Africa, that occurred between 60,000 and more than 160,000 years ago. It is quite likely that there were

---

* Like radiocarbon dating, U-series is an isotopic decay method but, unlike radiocarbon, the half-lives of some of these isotopes is much longer, so much older material can be dated – up to 500,000 years. The method focuses on the isotopes of uranium, beginning with ^{238}U, which decays to ^{234}U and then to ^{240}Th. Eventually the decay chain cascades down through several daughter isotopes ending in stable lead. ^{234}U and ^{240}Th are the most commonly used isotopes in the dating process. Uranium is soluble in water while thorium is not and this means that after uranium has entered the bone or tooth thorium will begin to accumulate; the more ^{240}Th, the older that specimen will be. When we date a tooth or a bone what we are in fact dating with this method is the time that has elapsed since uranium has moved into the material. Of course, this might happen a long time after that material enters the archaeological record, and so for this reason uranium-series dates of teeth and bones are usually considered 'minimum ages'. One of the assumptions with the method is that the sample is a 'closed system', which means no uranium has entered or left the material being dated since the original uptake of uranium. There are various methods that can be used to test this. In dating stalagmites or stalactites (or speleothems) the system is usually closed, but in bone and teeth this is less common.

other movements as well, and that future archaeological work will reveal them. New research in hitherto underexplored regions, such as the Arabian Peninsula, is beginning to produce exciting evidence.

There appear to be two possible pathways that modern humans might have followed out of Africa. The Sinai Peninsula is one. In a few days one can walk across the Sinai into the Levant, the area broadly encompassing modern Palestine, Lebanon, Jordan, Syria and Israel. Although we imagine the Levant to be part of Asia it is, in biogeographical terms, an extension of northern Africa. It is interesting to remember that a panoply of African or Afro-Arabian animals called this region home until comparatively recently. A research project I have been involved with has identified quite recent evidence for hyena, leopard, lion, camel, zebra and gazelle in Israeli caves, so it is probably not a surprise that we find early modern humans in the region as well, because if animals were able to move outwards it follows that humans could too. I have often noticed that a good indicator of human habitation is the presence of their prey; humans tend to place themselves in locations where they can take game, either by following them or ambushing them on migration routes.

A second path from Africa, which has received a great deal of recent attention amongst scholars, is the Bab al-Mandab strait, in the very far south of the Arabian Peninsula.* This strait would have been 5–15km wide at times of lowered sea levels, so would have required a maritime crossing. It is possible to look across the strait and see the far coastline, so human groups would know that new lands lay close at hand. Recent work in Arabia has shown that, far from being the dry desert of popular imagination, the region was periodically 'green' over the last 150,000 years, with lakes and rivers that formed as monsoon-influenced climate cycles moved north.[29] Analysis of satellite images has revealed that a patchwork of ancient rivers and palaeo-lakes once existed across the Arabian Peninsula.[30] This has been confirmed by ground surveys and later excavation. It appears to

---

* The strait lies between the modern countries of Djibouti on the African side and Yemen on the Arabian.

have been quite possible for both humans and animals to move across these regions in ancient times. It is a similar story in the great Sahara Desert. The presence of Nile crocodiles at desert oases and sub-Saharan African plants and aquatic fish found in the far north of Africa shows that wetter climates in the past opened up these now desolate environments to animal and human movements. Archaeologists have found the bones of hippopotamus, giraffe, elephant, frogs and lions in now parched and desolate deserts. The record of rock art from Holocene sites in northern Africa confirms that these wetter conditions have only recently ended. One sees animals depicted that require significant water sources, now no longer present. It is almost as if the artists are sending us a message, telling us how different life used to be in these regions.

New archaeological work has shown that humans were present in the Arabian Peninsula as early as 85,000 years ago, much earlier than the orthodox 50–60,000-year Out of Africa model would predict.[31] The Bab al-Mandab could have been the point from which humans moved into and across the Arabian Peninsula, and from there outwards towards western and eastern Eurasia.

How can we assess which of these two alternative pathways, the Sinai or the Bab al-Mandab, was more likely? Climate science has progressed substantially in the last couple of decades as scientists try to reconstruct past global climates in order to understand where the Earth's climate might find itself in the future. General Circulation Models (GCMs) are numerical models that seek to explain the Earth's climate system and they are now being increasingly used to try to reconstruct more precisely the climates of the deeper past. Researchers have explored the most optimal windows for movement out of Africa, focusing on one of the key indicators, rainfall. Hunter-gatherer groups cannot cope in environments where rainfall is less than 90mm per year; it is just not possible for them to survive. Climate reconstructions show that between 200,000 and 250,000 years ago, and again around 130,000 years ago, the Sinai route was crossable because reconstructed precipitation levels were higher than this, but that the Bab al-Mandab route was crossable for much more extended periods over the last 300,000 years. After 65,000 years ago the southern route was open and attractive for movement for more than 30,000 years.[32] It seems to me that the evidence supports the idea that both paths were probably used, but perhaps at different times, if the climate evidence is taken at face value.

If there were early exits of the age that are supported by the Greek Apidima humans and those at sites like Misliya Cave in Israel, then what happened to these pioneering groups? Would later modern humans coming out of Africa have met with others who had already been living outside of the continent for millennia?

Rather like the Israel situation we mentioned before, it is possible that the majority of these early dispersals ultimately failed, and indeed this is what some of the genetic evidence now seems to suggest. Large genetic studies using nuclear DNA sequences from modern people show that, in contrast to the 50–60,000-year-old estimates, the very early 120–130,000-year-old diaspora is less well supported. This could simply be because these early pioneers did not contribute significantly genetically to later human groups.[33] Two teams, however, have found hints of a small (~2 per cent) genetic contribution from a seemingly earlier diaspora in modern Papuan genomes.[34]★ In addition, there is some evidence for modern human DNA in a Neanderthal from Denisova Cave. Researchers have shown that some segments of DNA from the so-called 'Altai Neanderthal' (Denisova 5) derive from a modern human source that is closer to modern Africans. This genetic introgression is estimated to have happened around 100,000 years ago.[35] It is possible that this introgression happened in the Near East, where we already have evidence for the presence of early *Homo sapiens*, or it could have come from humans who had moved into more eastern parts of Eurasia. At present it is not possible to be sure. The groups that made it out to places like Apidima may have simply died out and left no subsequent record save for a small proportion of DNA buried deep in our own genes and recoverable today only by using powerful statistical tools.

Why did people move out of Africa? There are several interrelated reasons, some of which invite speculation because they revolve

---

★ These researchers called this xOoA (extinct Out of Africa) as opposed to OoA.

around human motivations, desires and actions that we can never truly know or reconstruct. I suspect that population size may be one factor. Group sizes are larger in hunter-gatherer groups in the tropics, compared with the temperate regions of the world. As groups get larger, so does the chance of them splitting and of a subgroup moving away.[36] Further areas of land are required for them to survive, and this results in expansion. Climate change and the tendency of humans to follow game probably also play a role. Climatic changes see previously marginal areas becoming attractive with swings to wetter conditions, and the reverse happens when climates become drier, hotter and less attractive. This can influence human groups to move and respond to 'pull' and 'push' factors. Populations can become fragmented in such periods of adverse climatic conditions. Climate change is the constant background to the early human story.

The humans who did leave Africa, regardless of when that was, would have encountered a range of new environments to adapt to, from the intense cold of Ice Age climates, the deserts and semi-arid landscapes of Central Asia, the warmer and wetter conditions of the tropical rainforests and, finally, that most challenging of barriers, the open sea. We know much more about the adaptation of humans to colder climes because the vast majority of excavation work has been focused on the more temperate parts of Eurasia. It is only comparatively recently that scholars have focused on dense tropical rainforests, which humans must have traversed early in order to reach places like Australia. This would have necessitated new ways of adapting to unfamiliar environments and significant new challenges in order to survive.

In 2003 I was working at Niah Great Cave in the rainforests on the island of Borneo (Malaysia), as part of a team led by Graeme Barker of Cambridge University. The archaeological site of Niah is located at the mouth of a vast and impressive cave system in part of a huge limestone massif, the Gunung Subis, around 15km from the modern shores of the South China Sea. It was discovered by the remarkable Tom Harrisson in the 1950s. He excavated there with his wife Barbara in the 1950s and 60s.

Tom Harrisson is about as close to a real Indiana Jones-like figure as you can possibly get in archaeology. As part of the British and Commonwealth war effort in the jungles of Malaysia during the Second World War he worked behind enemy lines, where he coordinated Dayak head-hunters armed with lethal blow-darts to neutralize the occupying Japanese forces. More than 1,500 soldiers were killed or captured. Later he became the curator of the Sarawak Museum in Kuching and pioneered conservation efforts for orangutans and endangered turtles.[37]

In February 1958 the Harrissons made their most famous discovery, the skull of an anatomically modern human – the so-called 'Deep Skull'. I worked on the dating of the skull and the archaeological sediments that had been excavated. We found that the Deep Skull was 42–44,000 years old, making it one of the earliest *Homo sapiens* remains outside Africa. The wider Niah team aimed to explore how it was that humans had managed to successfully colonize the rainforest biome. How did these early humans survive and adapt? Microscopic shards of charcoal, pollen and starches recovered from sediments in the cave revealed that, rather than solely wet lowland rainforests, these groups had to cope with a range of different environments. There was evidence for montane forest trees, savannah and grasslands, for example. It appears likely that the prehistoric occupants played a role in opening up and controlling the forest with fire.[38] They exploited and hunted leaf-eating monkeys and macaques, as well as pigs and monitor lizards. To do this the Niah people almost certainly used traps, probably by leg-snaring, and perhaps darts or arrows. They hunted orangutans, tortoises and turtles, collected freshwater shellfish, fruit, roots, tubers and the pith of sago palms and nuts from the forest. Large intercut pits that were excavated at Niah may have been used to process plants that have highly toxic acids, as modern people in the rainforest do today.[39]

Modern humans have been found in other rainforest contexts too. In Sri Lanka, for example, we find compelling evidence for the use of bow-and-arrow technology from as early as 48,000 years ago, as well as the probable use of poison on projectile points made of monkey bones.[40] We find evidence for toolkits that were probably used to make plant-based cords and to work animal skins. This suggests that items such as nets might have been used by the people living there. We normally associate early evidence for clothing with survival in cold climes, but in the rainforest it is possible that it could have been adopted to protect people from the range of insect-borne tropical diseases.[41] In Sumatra there is evidence for the presence of modern humans 65,000 years ago,[42] so it is becoming more and more likely that our modern human ancestors were well adapted to these new environments at an extraordinarily early date. Yet many researchers had thought these regions could not be occupied until much, much later.

In fact, we find modern humans in a variety of places from early dates, not only in the rainforests in Southeast and South Asia, but near the shores of the Mediterranean, in Australia and in the colder climes of Siberia. In attempting to understand this breadth of occupation, some researchers have coined the term 'generalist specialist' to describe the way in which these early humans occupied a diversity of environments, while at the same time appearing to adapt to extremely specific habitats.[43]

The evidence, then, suggests that some humans left Africa earlier than the traditional Out of Africa II model has suggested, certainly earlier than 120,000 years ago, and that there were probably multiple exits.[44] The early exit did not result in a wide dispersal of human groups across Eurasia and into Australia. It may have ended in extinction and disappearance. It seems that only after about 50–60,000 years ago do we find humans spreading widely across all parts of Eurasia, Southeast Asia and greater Australia.

One reason why humans failed to settle more permanently might be competition. Perhaps the presence of Neanderthals and other human populations in number was a key reason behind this. I think this might be one factor, but not the single explanation. I think it is more likely that before dispersing into the wide range of environments in Eurasia *Homo sapiens* had first to develop a gamut of new and improved technologies in order to survive and prosper. In the colder regions of Eurasia in particular, warm clothing, made of softened skins and furs, shoes or fur-lined boots, and perhaps even skis would have been required. New hunting technologies would have been needed to adapt to difficult new environments, like rainforests; from newer types of stone-tipped projectiles, capable of greater destructive power, to bows and arrows,[45] traps, snares and nets.[46] Without these items, life in the more extreme environments of the world would have been very challenging, if not impossible. Once developed, however, the archaeological evidence shows that modern humans could adapt and prosper. We see humans from this time onwards behave rather like an invasive species,[47] moving into a wide array of new environments, across the broad and varied expanses of Eurasia. Tracking these people across time and space has been extremely challenging, but we are steadily getting closer to reconstructing what happened. Groups of *Homo sapiens* who left Africa eventually met people with whom they shared a common ancestry, but had been separated for hundreds of thousands of years. And the best known of these are the Neanderthals.

3.   Neanderthals Emerge into the Light

On a late summer day in August 1856, in a valley near Düsseldorf in Germany, the story of human origins was changed for ever by a chance discovery. The valley, or 'thal', had been the focus of limestone quarrying for many years and the team of miners working there had almost exhausted the local source. One final area remained to be blasted: a large cliff face with two small caves situated twenty metres above the river. The caves were very difficult to get to from below, which is why they had been left until last. The miners scrambled down the sheer cliff from the top of the valley and laid charges inside the caves, intending to blow them up to start the mining

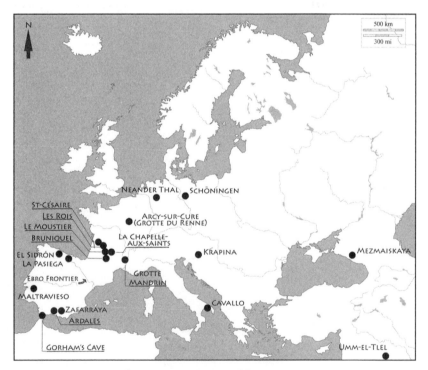

Figure 5 European and Near Eastern sites and locations.

process. Prior to this they went in to shovel out any sediment and broken rocks. The fill of the cave began to be rapidly and unceremoniously removed. It was then that one of them noticed something unusual protruding from the dirt.

A skullcap appeared, then some long bones, two femurs and some of the bones of an arm, a broken piece of pelvis and a few ribs. The remains were of very little interest to the miners, whose focus was on the limestone they were paid to extract, so they simply threw the bits of bone out of the cave down onto the valley floor. Later, the majority of them were collected by the quarry owner, Friedrich Wilhelm Pieper, who kept them in a wooden box. Pieper had a friend, a local teacher called Johann Karl Fuhlrott, who was interested in natural history. Pieper often called on him when something interesting turned up in the valley. He showed the bones in the box to Fuhlrott, who immediately recognized them as human. This horrified Pieper, who was concerned that the remains could be recent, perhaps of a murder victim. Fuhlrott reassured him that the bones seemed too old to concern the police.

The skull was human-like, aside from its extraordinarily thick brow ridges, which were far bigger than those of any modern person. Later, Fuhlrott took them to Hermann Schaaffhausen, a natural scientist at the Bonn Museum and it was he who described them to the world in a publication in 1858.[1]

The question of the age of the fossil remains was key in the minds of other scholars who considered the bones and what they were. Two years later, in an effort to determine more details about the precise location of the bones and their context, Fuhlrott went to interview two of the miners working in the cave that day, Luigi and Alessandro. Rather ill-advisedly, he turned up at the site with a lawyer and with neither of the usual incentives (money and beer). Sadly, perhaps because of this, the two claimed to have absolutely no recollection of anything at all.[2]★

★ Interestingly, 139 years later in 1997, German archaeologists led by Ralf Schmitz undertook an excavation in the approximate area of the former base of the cave. They recovered the remains of bones that, incredibly, included a fragment of the face and a piece of femur from the same specimen. They could be combined with the other bones rather like lost jigsaw pieces. They also found arm bone remains that indicated there was more than one individual buried in the cave. See note 2.

The valley is called the Neander Thal and the remains, of course, belonged to that most famous of 'other' human groups; the Neanderthals. The skeleton is the 'type specimen' or holotype of the Neanderthals; Neanderthal 1 or the Feldhofer specimen.[3] Taxonomically, we call Neanderthals *Homo neanderthalensis*.

It is apt, I think, that the literal translation of the Greek word *Neander* is 'New Man'. That a different type of human, a New Man, had once lived in Europe came as a profound shock to science. At the time the race-based approach to classifying human groups on appearance, inferred behaviour and intelligence had resulted, not surprisingly, in a grossly distorted and wholly false view of humanity. White Europeans (male) were viewed as being superior to all others. Africans and Australian Aboriginals were considered as primitive at best, and 'savages' by many. It was natural that these newly discovered Neanderthals should fit into the 'other' category too.

However, another finding, made in France in 1909 – the almost complete skeleton of the Old Man of La Chapelle-aux-Saints – demonstrated that, at least postcranially, Neanderthals were much more similar to us than to the traditional cavemen of popular culture.[4] Despite a reconstruction that depicted the Old Man (in fact, a man no older than forty-five) as brutish, squat and backward, there was something about him that drew attention. I recalled this in 2006 when I found myself in Paris, at the Musée de l'Homme, to take samples for dating from the famous Cro-Magnon modern human skeletons. During the course of my visit, the curator of their vast human collection asked whether I would like a look around the museum basement, where the most precious specimens were kept. Of course, I jumped at the chance. We travelled down and down in a small dusty service lift and found ourselves amongst cabinets and boxes, most of them securely padlocked. Gradually, one by one, out spilled the crown jewels of French prehistory: skulls from every age, beautiful Venus figurines, and then, to my enormous surprise and delight, the skull of the Old Man himself. Looking closely as I held it in my hands, I could see what others had noted before: that virtually all of his teeth were missing, lost before, rather than after, death. He must have had great difficulty gathering and processing food and surely would have needed the help of a community or wider group. He could not have survived alone. The Old Man of La Chapelle showed that there were more similarities than differences between us and the Neanderthals in terms of the comfort of a social web and network of support. But just how similar were Neanderthals to us more generally? Are they our evolutionary ancestors, or are we entirely separate? What contact did we have with them if the latter? Why are there no Neanderthals any more? These are questions that researchers have been debating, often bitterly, for more than 150 years.

As we have seen, we share a common ancestor with Neanderthals around 530,000 years ago, but we parted after this point and Neanderthals evolved separately and mainly in the colder climes of northern Eurasia, while we occupied Africa for much of our evolutionary history. Neanderthals, then, are our evolutionary cousins, rather than our ancestors. Much later, Neanderthals disappear from the fossil record in Europe at almost exactly the same time as modern humans appear to have arrived. This has been taken to imply a probable causation, suggesting to many researchers that we were probably superior cognitively and technologically compared with them. Neanderthals have therefore been considered for much of the last 150 years to be evolutionary dead-ends. This narrative of our superiority fitted because we survived, and they did not.

Research undertaken over the last few decades, however, has shown that this is an erroneous interpretation. The place of Neanderthals in our evolutionary story has changed dramatically in the light of new science and archaeological discoveries, improved excavation and the careful analysis of old museum collections. It turns out that we have seriously underestimated Neanderthals; they were much more similar to us, culturally, socially and technologically, than we ever thought.

Let's take subsistence and diet as an example. For many years it was suggested that Neanderthal diets were narrow and focused predominantly on the meat of large to medium-sized herbivores, lacking the wider dietary breadth that seems to have accompanied the arrival of *Homo sapiens*. Evidence in support of the importance of meat was taken from archaeological sites, but crucially too from evidence obtained from human bone collagen carbon and nitrogen isotopes, which suggested that Neanderthals were carnivores at the very top of the food chain and therefore predominantly meat eaters.[5] Despite this emphasis on meat, some suggested that Neanderthals were more like scavengers than capable hunters.

The common interpretation of Neanderthals as a cold-adapted species seemingly confirmed these suggestions, since other human hunter-gatherer groups living today in colder, Arctic environments show a similar dependence on meat. This diet has been argued to be stable over time, with little in the way of change or variety. Interestingly, these data have been used to infer aspects of Neanderthal social structure. Some researchers, for example, have suggested that Neanderthals probably did not have a sexual division of labour, as often seen in hunter-gatherer groups, where men hunt animals and women more often collect and process plants, tubers and so on. Instead, the high reliance of Neanderthals on meat has been taken to suggest that both sexes were probably involved in hunting or scavenging. From this it has been further suggested that modern humans had an important advantage over them because *Homo sapiens* women could procure a wider range of foods. This ignores the fact that groups that are reliant on meat today, such as the Inuit, also have a sexual division of labour.*

It is important to note, however, that Neanderthals lived across a vast geographic range within which there were variable climatic and environmental regimes, so it is perhaps not unexpected that we would see variations in diet across space too. This is a key point, because it turns out that the proportion of meat versus plant material in diets is strongly linked to climate. In warmer climates, hunter-gatherers tend to exploit more and wider ranging plant foods.[6] In addition, no human group can survive on meat alone; they must surely have eaten plant foods occasionally.

Recent work has shown that they did, and that their diets had far

---

* Women in Inuit communities are usually involved in hide tanning, cooking, heating houses and cleaning, while men are involved with hunting, house-building and fishing.

more variation than we previously thought. In southern Spain, for example, we have evidence for marine resource use amongst Neanderthals, including the hunting and catching of seals, dolphins, fish and shellfish.[7] At sites in Greece and Italy there is evidence of turtle and bird hunting, as well as shell-fishing.[8] Microscopic evidence contained in dental plaque from Neanderthal teeth has yielded evidence for starches and phytoliths from plants, along with traces of seeds, fruits, nuts and vegetables.[9] Neanderthals appear to have eaten lentils, water lilies, pistachio, tubers, wild cereals, figs, mushrooms, pine nuts and mosses, and many more unexpected types of food.[10] They were certainly not exclusively meat eaters. I think our previous interpretations have been influenced by the overwhelming presence of animal bones in Neanderthal archaeological sites and the lack of other information about dietary sources, mostly because plant and vegetal material does not survive well in the archaeological record.

What else did Neanderthals consume? At the El Sidrón site in northern Spain archaeologists have found somewhat unsavoury

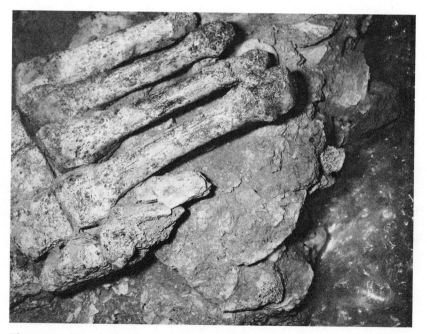

Figure 6 A Neanderthal foot emerges from El Sidrón.

evidence of a gruesome event that took place 49,000 years ago. In 1994 cavers there discovered a chamber in an underground complex that contained some human bones. It was initially assumed that the remains might belong to victims of the Spanish Civil War. In fact, they were much older. Bones from around thirteen Neanderthals were eventually excavated.[11]* The dig revealed that the bones were redeposited together from somewhere else in the cavern, but essentially were the result of a single catastrophic event; they all died together – men, women and children. But how? Careful analysis of the bones showed that they were literally covered in hundreds of small cuts caused by stone tools, with other bones broken and snapped. I remember analysing the bones for dating and isotope work and seeing bone after bone with evidence for cutting, gnawing and breaking. I saw the bones of a whole Neanderthal foot still encased in its sediment after having been thrown aside. The Neanderthals of El Sidrón had been butchered and consumed, judging by the date almost certainly by other Neanderthals.

Was it the result of hunger? Was this ritual behaviour? We can only speculate. It does appear, however, that cannibalism was not uncommon amongst Neanderthals generally. There are many examples of butchery marks on human remains and evidence shows that human bones were treated similarly to the bones from other prey species.[12] I think that what this shows is that Neanderthals might have had enemies and other groups or bands that they killed and, sometimes, ate. It is tempting to speculate further that this might have been occasioned by fighting for access to better hunting areas or cave sites, but there is little compelling evidence in support of this.

In terms of comparing modern *Homo sapiens* and Neanderthals, we can see some similarities in their adaptive strategies; they appear not to be so different from each other. It seems to me that dietary variation, if it exists, might be influenced more by the environment and what resources were available locally than by behaviour or foraging

* Seven adults (four females, three males), three teenagers (one female, two males), two juveniles (one male, one whose sex is unknown) and one infant (sex unknown).

ability. In terms of daily subsistence, it seems that Neanderthals 'eat what is there',[13] focusing on the resources to hand, including those that are seasonally restricted. This shows that a degree of planning and organization must have been needed, giving us a glimpse into Neanderthal social organization and communication.

This growing evidence of a more complex adaptation is also illustrated by new data that show that Neanderthals may have used plants to treat the sick. This information has been derived from biomolecules archived in tooth plaque. One of the Neanderthals from El Sidrón had a tooth abscess. Molecular compounds from yarrow and chamomile were identified in the subject's dental plaque. These bitter-tasting, low nutritional value plants strongly suggest the possibility that they might have been used as medication.[14] Ancient DNA and proteins extracted from the same dental plaque yielded evidence for salicylic acid. This is the active ingredient in aspirin, derived from poplar trees. I wonder whether the Old Man of La Chapelle was treated in such a way. I hope so. The El Sidrón individual probably also suffered from acute diarrhoea caused by a bacterium. Sequences of DNA derived from a *Penicillium* mould, the active source of penicillin, again hints at self-medication using natural antibiotics.[15]

Other aspects of Neanderthal adaptation and behaviour have also come under renewed scrutiny over the last decade. One of the most visible aspects of the archaeological record are stone tools. European Neanderthals made a stone-tool industry we call the Mousterian, named after the site of Le Moustier in France. It dates from the Middle Palaeolithic part of the Stone Age and is characterized by a method of making implements called the Levallois technique, dating from about 250,000 to 300,000 years ago. Levallois is a sophisticated method of making beautiful curved flaked tools, and is named after a suburb in the west of Paris near the Seine, where this type of tool was first found. A piece of stone is first fashioned into the intended shape of the tool; this is the so-called 'core'. One side of it is flaked so that overall it resembles the shape of a tortoise shell. From the base of this shell-shaped core, using a single blow from a so-called striking platform, a curved flake tool can be produced. Levallois tools are characterized by a substantial increase in the size of the cutting blade compared with earlier tools. We do not really know for sure where or who first started to fashion Levallois implements – they have been found in African sites back almost 300,000 years – but Neanderthals are the group with which they are most commonly associated.* Some have suggested that because there is a general lack of major cultural change in the lithic toolkits of Neanderthals, they lacked the capacity for innovation. The alternative argument, of course, is that the toolkit was adequate for all of their needs and was made to last.

It turns out, however, that Neanderthals were much more innovative than thought, and did use other, more complex forms of technology to make tools. Some 72,000 years ago, at the site of Umm-el-Tlel in Syria, Neanderthals transported small balls of bitumen to heat for hafting their Levallois weapons, probably to wooden spears.[16] Evidence from a range of other sites shows that Neanderthals used tar from birch bark to provide handles or for hafting tools. We know that Neanderthals also made *lissoirs*, or bone smoothers, that may have been the key tool for softening leather and working hides.[17] Previously these were considered the sole preserve of *Homo sapiens*. They are not, and in fact it is possible that moderns learnt to work hides like this from the Neanderthals.

Other types of material that could have been used to make tools, particularly wood, do not survive nearly as well in the archaeological record but in some lucky cases we have evidence for these more perishable items. In the 1990s at the German site of Schöningen, for example, the potential importance of wood was brought into sharp focus with the discovery of nine beautifully made spears, a lance, a burnt stick and a double pointed implement, found amongst the remains of butchered horses dating to about 300,000 years ago.[18] The implements were originally deposited near an ancient lake shore and preserved in sediments before being revealed during open-cast lignite mining. They may have been made by *Homo heidelbergensis* or another, earlier related hominin rather than by Neanderthals, but

---

* *Homo sapiens* also used the method to make Mousterian tools in other parts of the world, such as the Levant, so in different regions they are not the exclusive hallmark of Neanderthals. In Europe, however, the Mousterian does seem to be exclusively Neanderthal.

they suggest to me that such technology was almost certainly used from this date by our human family. One wonders how many more examples of such technology have failed to survive the ravages of time and therefore can never be found by archaeologists.

Food and tool choices, however, are not necessarily indicative of complex cognitive behaviours. Were Neanderthals able to express themselves in abstract and symbolic ways? Could they make art, recall their dreams or joke? And, if there was a difference between us and them, could this be the reason for our ultimate success in survival and their demise?

Archaeologists have to infer behaviour from the record of materials and artefacts, which is tricky. At the dawn of the Early Upper Palaeolithic in Europe, around 45,000 years ago, we see a proliferation of objects in the archaeological record such as perforated teeth and shell pendants, the use of pigments and colourants, decorated and incised bones, carved figurative art and cave painting. This occurs at about the time *Homo sapiens* first arrived. Objects and ornaments such as these are important because they give us an insight into the human mind, and particularly what we call behavioural or cognitive complexity. Complex symbolic objects allow information to be transmitted to others, sometimes over great distances. Symbols are powerful; they can be used to bind people together, to confer a degree of membership, to reflect rank or seniority or bring people from afar into a joint network or alliance. The benefits of this type of system are far-reaching. In times of economic or subsistence stress, for example, they may confer an advantage to some groups and not others. A shared identity and group alliance make it more likely that help will be given in times of need and that others can be relied upon to provide assistance. Networks and links between groups can also reduce dangerous inbreeding in small hunter-gatherer groups and ensure that genetic diversity is maintained. In addition, maintaining long-distance connections and social networks can help to retain and spread cultural innovations that might otherwise disappear when a small group goes extinct. This multilevel sociality has been implicated as one of the prime movers of the successful migration and dispersal of low-density human hunter-gatherer groups around the

world and their more effective transmission of ideas and knowledge
to one another through time.[19]

The suddenness with which this explosion of different types of symbolic artefacts seems to have appeared at the time of the Early Upper
Palaeolithic in Europe has been taken as evidence of the relatively rapid
arrival and dispersal of modern humans and the concomitant demise of
Neanderthals. It has often been assumed that Upper Palaeolithic objects
are associated only with modern humans. But were they *exclusively*
associated with us? Increasingly, there is evidence that suggests not.
The inference therefore is that *Homo sapiens* were not the only ones capable of such advanced behaviour.

At a Neanderthal site in the south of Spain, for example, archaeologists found shells with perforations that enabled them to be worn
or displayed. In addition, the remains of pigments were found on
and near the shells.[20] These included ochre (red), pyrite (black) and
natrojarosite (yellow) minerals. A small bone with pigment present at
its tip implies its use in preparing these pigments and mixing them to
make different colours, mixed in a larger *Spondylus* shell container.
The remains date to around 115–120,000 years ago,[21] well before
Upper Palaeolithic modern humans are present in Europe. In ancient
Egypt natrojarosite was used as a cosmetic. Could Neanderthals have
been using these various pigments to the same effect or for self-
decoration? I think the answer must be yes. In South Africa, at the
site of Blombos Cave, similar but better-preserved evidence for
pigments and colourant preparation is taken to reflect behavioural
complexity amongst modern humans.[22] Surely, the same must apply
to the Neanderthals of Spain.

In addition to shells, Neanderthals seem to have had a genuine
interest in the feathers of big birds of prey. They may have used them
in a decorative and therefore symbolic manner. This evidence is based
on the presence in archaeological sites of certain birds' wing bones
that have been clearly cut and sawn by stone tools. The argument is
that, since these bones derive from the wing regions and are not
nutritionally valuable, there must have been another reason to focus
on them. It might be that they were targeting the wings for feathers,
and perhaps these were being used for decorative purposes. There is

now a total of sixteen Neanderthal sites that have furnished examples of bird bones linked with this behaviour.[23] There is also increasing evidence that Neanderthals used eagle talons as well. Talons which have been deliberately cut with flint tools have been identified at sites in France[24] and Croatia.[25] At the Croatian site of Krapina there is evidence for talons being worn with a cord, perhaps around the neck.[26]

In popular culture we often associate Neanderthals with caves. While it is true that the majority of evidence associated with this group is found in caves, this is usually present at the front of the cave rather than deep inside, presumably due to the cold inside caves and the preference for camping at the mouth to maintain visibility. In February 1990, a fifteen-year-old boy called Bruno Kowalsczewski made a remarkable discovery that changed this presumption, and also changed our understanding of aspects of the Neanderthal mind.[27]

Bruno's father had long wondered about the origin of tiny puffs of wind that he had felt coming from behind a scree slope in the Aveyron region of France at a site called Bruniquel. Speleologists sometimes find lost cave entrances by feeling air currents like this. In winter the air is warmer inside a cave than outside, so puffs of warm air emerging from behind scree can be important indicators that a cave entrance might lie sealed behind the rubble of an ancient landslip. Inspired by his father's hunch, Bruno dug into the scree; for three years he tunnelled slowly, more and more deeply into it. After thirty metres of digging, the tunnel he had created finally opened up and he was able to squeeze, along with some of his caver friends, into a wider interior chamber.[28] He had found a cave that had been sealed off for millennia. Inside he could see stalagmites and stalactites, as well as evidence for ancient occupation by bears. Some of their bones lay on the floor of the cave, along with evidence for hibernation nests and scratch marks on the walls from their claws as they woke up from winter slumber. Further inside, 336m from the entrance, however, they found something that could only have been made by humans. Broken stalagmites, some burnt, were laid deliberately into shapes. Two large circles had been formed, one about 7m across and a smaller one 2m in diameter. In some areas the stalagmites were stacked one on top of the other.

There were traces of fire all around. Who had made the strange features, and when?

Dates later obtained indicated they were more than 170,000 years old. They were almost certainly, therefore, made by Neanderthals. Whether Bruniquel was a meeting place, somewhere for shamans to commune with spirits, or perhaps a location for funerary activity, we cannot tell. Until this discovery we had no evidence to support Neanderthals being habitually deep inside caves, we simply assumed that they were not able to, afraid to, or simply not interested enough to go there. Yet in Bruniquel we have evidence for the building of structures, the use of fire and the possible gathering of people more than 170,000 years ago, long before we find compelling evidence for the presence of *Homo sapiens* on the continent.

Neanderthals did not just enter these deep caves; recent work suggests that they may have also made abstract art and rock paintings while there. Once more, this was previously seen as the exclusive preserve of modern *Homo sapiens*. In Gibraltar, at the site of Gorham's Cave, evidence for possible abstract art was revealed in 2014, in the form of a so-called 'hashtag' image carved into rock at a date that means it could only have been made by Neanderthals.[29] Once more, we are mystified at what this was for or what the author or authors intended to communicate by it. At other caves in Spain, such as at Ardales, Maltravieso and La Pasiega, painted art on walls previously assumed to be the work of modern humans has produced surprisingly old ages.[30] The art included red markings, as well as a human hand stencil, that most evocative of representations. Tiny calcium carbonate growths found covering these artworks like a film can be dated using uranium-series (U-series) dating. This showed that the oldest art dated to *at least* 65,000 years ago, well before the earliest evidence for modern humans in Europe and their associated art. U-series dating provides minimum ages, so the age of the art is older than 65,000 years, but by how much, we do not know. Could Neanderthals have been responsible for painting more of the caves we know about in France and Spain? We need to obtain more direct dates from the artworks themselves to be sure. If the latest dates are reliable then the evidence suggests that Neanderthals, not *Homo sapiens*, were the first cave painters. Who knows, modern humans might have seen these early representations and copied them?

Increasingly, then, the weight of evidence is swinging towards Neanderthals not as ignorant cave-dwelling brutes, but as much more similar to us than we previously suspected. This shift is recent. It also raises the distinct possibility that the first *Homo sapiens* in Europe learnt new things from Neanderthals.[31] Many researchers have thought that any cultural connection would have been one way; from us to them. For this to have happened, of course, we would have needed to have actually met. Whether we did or not has been one of the most bitterly debated questions in Palaeolithic archaeology.

Archaeologists in Europe have mapped a series of archaeological sites and stone-tool industries dating to between 40,000 and 50,000 years ago that document the stirrings of innovation. We see evidence for increasingly small stone tools, so-called bladelets that can be just millimetres in length. We do not really know their intended purpose, but based on our understanding of later examples they were probably used for making projectiles, such as composit spearpoints or arrowheads. We see evidence for blades of a larger size being used as well. Increasing numbers of bone points appear, along with hints of the personal ornaments and symbolic objects we have already seen. Gradually, as we reach 40–42,000 years ago, these objects begin to build in number, and are associated with archaeological industries known as the Aurignacian that we link with incoming modern humans. Here, at dozens of sites across Europe, a range of new implements make their appearance for the first time. Ingenious points made of bone and antler that have a carefully manufactured split base, allowing them to be hafted to spear points much more efficiently; tiny bladelets made with an entirely new and distinctive production method, and a range of decorative and ornamental items. Large numbers of personal adornments make their appearance; over 150 different types have been found, made of shell (sixteen different species), antler, bone, ivory and, occasionally, even fossils such as ammonites and belemnites, along with tooth ornaments or pendants from a wide variety of animals, including wolf, hyena, reindeer, lion and even human.[32]

Slightly earlier than the Aurignacian, however, before 42,000 years ago, we see other stone-tool industries, which we term 'Transitional' (meaning transitional between the Middle and Early Upper Palaeolithic). The makers of these artefacts have been very difficult to identify. Debate has raged about whether the industries arrived with incoming *Homo sapiens* or were developed locally by late Neanderthals, perhaps influenced by encounters with modern human groups. The answer is key to understanding the process of the replacement of Neanderthals by modern humans and how this might have happened. Some of the stone-tool industries have a strong regional character, for example the Châtelperronian of northern Spain and France, the Uluzzian of Italy and Greece and the Bohunician of central and eastern Europe.*

The Châtelperronian has been associated with Neanderthals based on evidence obtained principally at one key site: the Grotte du Renne at Arcy-sur-Cure, south-east of Paris. In the 1950s and 60s, Mousterian archaeological levels were excavated there, above which were three layers that have become known as Châtelperronian. This industry is characterized by a desire on the part of its makers to produce blades, with one of the tools, the eponymous knife, being the characteristic type. Twenty-nine Neanderthal teeth and a tiny ear bone were found within the Châtelperronian levels of Arcy. Archaeologists also found pierced teeth of animals such as fox, deer, hyena, bear and horse and curious decorated ivories in the form of rings in the same levels – a range of artefacts similar to those we usually see in the Aurignacian, made by *Homo sapiens*. Then, in 1979, at the French site of Saint-Césaire, a Neanderthal skeleton was excavated, and the archaeologists there associated it with the Châtelperronian level. Since Neanderthal remains are found together with symbolically laden ornaments and objects, this must mean that Neanderthals made

---

* There is another called the Lincombian-Ranisian-Jerzmanowician, mercifully shortened to LRJ. The region in which these tools are found extends for 1,500km, from Britain, the Low Countries and Germany all the way to Poland, where, between latitudes 50 and 55° N we find sites containing very similar 'leaf-point'-shaped hand axes. We remain uncertain of whether this industry was made by Neanderthals or by us.

them, implying once more that Neanderthals too had symbolic or complex behaviour.

Not everyone is convinced, however, that the association between Neanderthals and ornaments at Arcy is legitimate. I have been one of them. It does seem curious that of the hundred or so Châtelperronian sites, only Arcy and one other* have any evidence for ornaments. If this were common amongst Neanderthals, surely there ought to be more. Others have suggested that the evidence at Arcy could be the result of mixing in the site: ornaments made by modern humans after the end of the Châtelperronian might have slowly moved down the sequence into the Châtelperronian layers. Or perhaps Neanderthals simply copied *Homo sapiens*' examples.

Dating the site and its objects is clearly crucial. I worked at Arcy back in 2006. I wanted to test the question of whether the stratigraphic sequence was reliable and in order, with respect to depth. Usually, as we go from the lowest archaeological layer in a site towards the top, we expect the ages to grade from older to younger in line with the sequence of deposition.

The results we obtained were somewhat mixed: around 40 per cent of them were consistent with the age expected, but many were substantially younger. We concluded that the site must be partially mixed, as had already been suggested, and therefore that the association between Neanderthals and symbolic artefacts was questionable. Our interpretation was strongly critiqued.[33] It was suggested that our dates were the problem, not the site.†

---

* Quinçay, also in France.

† It was at about this time that my father asked me what I'd been working on lately. When I replied a site called Arcy, my dad's face lit up. Arcy-sur-Cure? The Grotte du Renne? He told me that he and my uncle Richard, at the ages of nineteen and twenty-one respectively, had travelled to the site to work there in the summer of 1958 with André Leroi-Gourhan and his team. I couldn't believe it and nor could he when I told him that it was now one of the most famous and important sites in all of Europe and key in debates about Neanderthals and modern humans. As a teenager, he'd been excavating the key levels I was working on. We talked a lot about his recollections and memories of the site, and he produced some photographs of his time there from the dusty family archives.

More radiocarbon dates were produced from Arcy by another team, this time seemingly in contrast to our own, with hardly any significant outliers, suggesting that the sequence was not as mixed as we had suggested.[34] The difference might be that, while we selected more of the artefacts to date, these could have been treated with chemical preservatives while the other samples were untreated bones. Some of our dates might be too young, owing to this potentially unremoved contamination.

There is also the key Saint-Césaire Neanderthal skeleton to consider. Some have called into question the association between it and the Châtelperronian layer in which it was found and said that it is impossible to make a confident link between the two.[35] Some scholars maintain that the association between artefacts and Neanderthal remains at the Grotte du Renne is sound; some reject the site as mixed. Like many debates in archaeology, this one rolls on and on.

If we accept for a moment that Neanderthal and symbolic ornaments are indeed commingled in the Châtelperronian at these two sites and there is no accidental mixing, how should we interpret the evidence? Some have argued that the sudden appearance of what are often characterized as modern human items within a Neanderthal site constitutes an 'impossible coincidence',[36] and that surely Neanderthals must have borrowed or copied the cultural traditions of body ornamentation after modern humans arrived in Europe. This has been termed the 'acculturation' hypothesis. Alongside this, the question was always asked whether it was possible that this acculturation process involved biological as well as cultural exchange. Did our ancestors interbreed with Neanderthals? One model of human origins posits that instead of modern humans replacing Neanderthals there was genetic exchange between the expanding modern population and the indigenous European Neanderthals, with the latter being assimilated into their communities.[37]

I think the evidence increasingly shows that Neanderthals were more capable than previously thought. This has made it much less likely that Neanderthals were 'acculturated' by modern humans. The weight of evidence now suggests that if there was cultural transmission it probably occurred in both directions, and that the earliest evidence for the beginnings of complex behaviour in Europe was prior to the widespread arrival of *Homo sapiens*. Dating is clearly crucial and we will hear more about this in later chapters. If cultural transmission and exchange happened, then did our ancestors do more than simply *meet* Neanderthals?

This question has captured the popular imagination as well as science, with several best-selling fictional accounts exploring the 'what if' scenarios.* The first mitochondrial DNA evidence extracted from Neanderthal bone remains, however, showed no signs of Neanderthal and *Homo sapiens* interbreeding.[38] This suggested that either there was little to no overlap or contact between the two, or there was a biological reason, an incompatibility that meant they could not interbreed and produce viable offspring. I remember the early mtDNA evidence confirming my hunch that there was no interbreeding and very little contact at all between the two groups. As we shall see throughout this book, this interpretation has been shown to be entirely and indisputably wrong.

To get to grips with this question, however, it was necessary to sequence the nuclear genome. In 2006 the Neanderthal Genome Project was begun by Svante Pääbo and his team, its aim being the reconstruction of the nuclear genome of the Neanderthals. Just four years later, in 2010, the major landmark paper from the project was published. A comparison of the human and Neanderthal genomes showed unequivocally that between 1 and 4 per cent of the DNA in people outside Africa comes from Neanderthals.[39] It was true: we *had* interbred with our Neanderthal cousins! The long-assumed scenario of replacement without interbreeding was overturned: Neanderthals live on partially in us.

* The most famous of these is probably the series of books by Jean M. Auel that starts with *The Clan of the Cave Bear*, in which a five-year-old *Homo sapiens* girl called Ayla is taken in by a band of Neanderthals after her family's camp is destroyed and she is orphaned. Some years later she has a child after being brutally raped by one of the Neanderthals. The child has features of both groups but is considered deformed. Ayla is eventually banished, leaving her child to grow up with the Neanderthals.

Why did the mitochondrial genome not show any evidence for Neanderthal introgression in us when the nuclear genome did?* One possibility could be that any hybrid offspring with Neanderthal mothers were raised in Neanderthal groups, and disappeared alongside Neanderthals when they became extinct. A second is that interbreeding caused problems for the carriers of Neanderthal mitochondrial DNA and meant they could not produce viable offspring. Finally, it could be that female Neanderthals and male modern humans could not produce offspring that were fertile. In the sex chromosomes in the nuclear genome there are indeed some indications of hybrid sterility. Modern human X chromosomes and DNA in the germline (testes) show a reduced Neanderthal ancestry, and this has been taken as evidence that hybrid offspring *may* have struggled to produce children of their own. Whatever the reason, it is clear thus far that the mitochondrial DNA evidence is silent about this particular interbreeding story.

Could we have predicted the interbreeding? Many now think we ought to have. It turns out that, amongst primates, hybridization has been observed between species that diverged more than four million years ago. We share a much more recent common ancestor with Neanderthals by comparison. As we will see later in this book, interbreeding and hybridization not only appear to have been common amongst the various members of the human family in the last 200,000 years, but also might well be the key to understanding how it was that we ultimately became such a successful global species.

The bitter debate about whether Neanderthals and modern humans met and interacted archaeologically has abated to a large degree principally because of the discovery of interbreeding. If our groups were interbreeding, then cultural transfer – the exchange of ideas, thoughts and language – may well also have been happening. Humans are good at picking up new ideas, at copying others and learning new tricks, particularly when survival in tough environments is

* See pages 70–71 for an explanation of the differences between the nuclear genome and mtDNA.

important. Whether there was conflict or relations were peaceful, we can only guess; there is little firm evidence. I suspect both were happening at different times and in different places. Modern humans have peacemakers and warmongers, doves and hawks, and Neanderthals were probably no different.

One site that offers intriguing insights is Les Rois in France. There, archaeologists excavating a site dating to the Aurignacian period found a number of human teeth and two human mandibles, one from a modern human and the other from a Neanderthal child.[40] The Neanderthal jaw had been cut by someone using a sharp stone tool. The marks were similar to those left after the processing of animal bones at the site for meat removal, such as reindeer. Was this Neanderthal child eaten by modern humans? It's difficult to be certain. In some Aurignacian sites we have evidence for an interest in using human remains for decorative purposes; there are cases of perforated human teeth, for example, probably worn as a necklace. In this case it is possible that the Neanderthal remains could have been used for such a purpose. Perhaps the child was used for both.

From the discovery of the first Neanderthal remains, for well over a century, palaeoanthropologists have debated and discussed the Neanderthals and our relationship with them. Whether we were directly related ancestrally, whether we ever met one another and whether we were implicated in their disappearance. Great progress has been made but debates continue. What we did not realize, however, was that Neanderthals were only part of our ancient human story; there was another human relative in Eurasia living at the same time as Neanderthals that we had absolutely no clue about. The world of human evolution was about to receive a huge shock. Enter the Denisovans.

4.   The Road to Denisova Cave

The story of the Denisovans begins with a solitary man in a cave in the Altai region of Siberian Russia. The Altai is about where Russia, China, Mongolia and Kazakhstan meet. According to local oral tradition, the man, a monk called Denis, had occupied the cave in the late eighteenth century. He had lived there for many years as an ascetic, giving his name to the cave we now call Denisova.[1] But he was not the first occupant of the cave. He lived on top of the archaeological

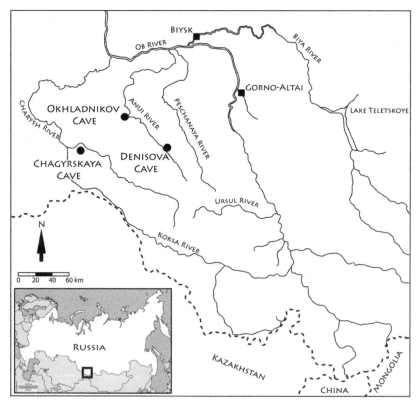

Figure 7 Sites and locations in the Altai.

remains of others who had lived there over the course of 300,000 years. This prior evidence would be revealed over a century later by a team of archaeologists from the Russian city of Novosibirsk.

Denisova is located around halfway across Eurasia in the region of Siberia. Siberia is vast. It stretches from the Ural Mountains in the west, forming the boundary between Europe and Asia, to the Pacific Ocean in the east, across eight time zones. Siberia includes the Far East territories along the Pacific Rim, up to Chukotka, which sits across the Bering Strait from Alaska. To the south, Siberia borders China and Mongolia, and the northern steppes of Kazakhstan. The vast empty lands of Yakutia* lie to the north, occupying a fifth of all Russian territory with its so-called Taiga, a zone of boreal forests and permafrost. Every few months, it seems, Yakutia reveals another frozen Ice Age mammoth or lion or wolf, complete with hair and teeth, blood and tusks, released from its steadily melting icy grip. To the north of the Taiga is tundra, with treeless, marshy plains, white nights and herds of reindeer. The Baikal and Transbaikal regions, with the eponymous lake said to hold an astonishing one-quarter of the world's fresh water, sit to the south and to the east of Denisova and the Altai region.

The Altai around Denisova is subalpine and dotted periodically with small, traditionally built wooden houses of the local Altaian people. Above, you can usually see the dark outline of large eagles soaring over river valleys hemmed in by high craggy peaks cloaked with pine, birch and spruce. Horses and pigs roam freely around the occasional village. It is said that the horses are from the same stock that furnished the steeds of the army of Genghis Khan. The air is beautifully clean and the river waters crystal clear and cold. Summer meadows are replete with wild berries and bright flowers of many colours. The late-nineteenth-century Russian poet, painter and writer Nikolai Rerikh called the Altai 'the Pearl of Asia', and one can see why after only a few hours there.

I first heard about the Altai from Paul Haesaerts, a Belgian friend and colleague, who told me it was one of the most beautiful places

---

* Now called the Republic of Sakha.

he'd ever seen and that I simply had to go there. I made a mental note that I must follow his advice one day. I am glad I did.

But Denisova is remote, so its secrets have remained hidden. Until quite recently, reaching the cave itself was a challenge. The roads that traversed the Altai were of poor quality and sometimes impassable, so archaeologists had to be dropped there by helicopter and camp for weeks or months at a time.

To reach the Altai and Denisova Cave I usually fly via Moscow to the city of Novosibirsk, the capital of Siberia and Russia's third largest metropolis. During night flights the terrain below is dark for long stretches, save for the very occasional twinkling lights from a village or small town. I always imagine how hard it must have been walking on foot during the prehistoric period across this vast and desolate space.

After reaching Novosibirsk we drive for five or six hours across the flat steppe of southern Siberia, through miles of wheat and sun-flower fields in the summer, before the topography changes and low hills come into view. The roads become rougher and more shingly, potholes appear, and the path is occasionally washed out completely by a river. Just as the bouncing and lurching of the four-wheel drive becomes intolerable, and after some eleven hours' travelling in total, the base camp of the Denisova team finally appears and it's time to see old friends, settle in, unpack and relax, usually over a few shots of vodka and an amiable dinner.

The first time I ate with my Russian colleagues, I had to explain to Professor Michael Shunkov, co-director of the excavations at Denis-ova, that I was a vegetarian. When his translator conveyed my message Michael immediately replied, in perfect Russian-English, 'You will not survive in Siberia!' Meat is indeed usually on the menu, but Russian hospitality means that I have never gone hungry yet.

The first Palaeolithic site in the Altai Mountains was found in 1961 by A. P. Okladnikov, the father of Siberian Palaeolithic archaeology. Later, the Siberian branch of the Russian Academy of Sciences in Novosibirsk started to build the range of accommodation and facilities that now mark the permanent Denisova base camp. It is a very pleasant place, with a range of chalet-like buildings, proper run-ning water and a nice dining room and kitchens, and recently even a spa and sauna. The contrast with the other excavations in the Altai could not be starker. At sites like Chagyrskaya and Strashnaya, around four hours by four-wheel drive from Denisova, visitors sleep in a tent and cook over an open fire. By comparison, the Denisova camp is luxurious. When I started visiting the site there was no mobile phone coverage to speak of, but by 2018 I and the research team were able to watch the football World Cup on a TV in one of the guest chalets.

The cave is about 100m from the base camp. It sits 28m above the Anui river, which bubbles constantly below on its long journey to the massive River Ob.* Eventually, the same water will reach the Arctic Ocean. The fame of the site spread rapidly after news of the discovery in 2010 of a new human species in the cave was reported widely in the international media. There is even a tiny hut just beneath the site selling fridge magnets and postcards, copies of artefacts and key rings. On the road there is now a huge red Денисова Пещéра (Denisova Cave) sign with an arrow pointing up the incline to the site. More and more Russians seem to be aware of the Denisovans and the importance of the cave in the world of archaeology. Slowly the region is opening up. It is exciting on the one hand but sad on the other, because it takes away a little of the wild romance, that wonderful feeling of adventure and of being lost and distant from the rest of the world.

Over most of the 10,000 years before Denis moved in, we have a great deal of evidence for earlier people. Descending the well-stratified archaeological sequence from the current surface down through layer upon layer of soil and sediment reveals the tell-tale signs of pottery vessels and artefacts of bone, stone and horn that tell us about the presence of people in the Middle Ages, the Huns and Sarmatians and the first people with knowledge of how to forge iron. Digging deeper, evidence for the use of bronze emerges. Earlier still, we see the arrival of the first farmers, changing the human way of life for ever. There were people here as humans went through millennia of crushing climatic lows during the so-called Last Glacial Maximum, from around

---

* The Ob is the world's seventh-longest river, more than 5,300km in length. Its watershed embraces the Altai and reaches the borders of China and Mongolia.

24,000 to 19,000 years ago, and 14,700 years ago, when the planet suddenly and dramatically warmed by 7–10°C in three years for reasons
not yet fully understood. There is also evidence of habitation long
before this. Some of the earliest members of our species, anatomically
modern humans or *Homo sapiens*, were probably living here perhaps as
long ago as 40–50,000 years, although the evidence is largely circum-
stantial at present, as we shall see in Chapter 10. Before this,
Neanderthals were present on several occasions. Since 2007 we have
known Neanderthals reached as far east as central Siberia, when
geneticists teased out their DNA from a previously unidentified femur
at a nearby site in the Altai named in honour of the aforementioned
Okladnikov, and we have some of their bones in deeper archaeological layers at Denisova itself. The first people at the cave, however,
were not Neanderthal or us. They were Denisovans.

What did Denisovans look like? What would we have seen in them
that would differentiate them from us and/or the Neanderthals?
Through new science, excavation and analysis, we are starting to
build a picture of these newly discovered people, our new cousins.
As we shall see, recent work has progressed to such an extent that it
has allowed us to reconstruct what a Denisovan might in fact have
looked like.

The first person to excavate at Denisova was the Russian palaeontologist Nikolai Ovodov, in 1977. He was drawn to the site by virtue
of the rarity of caves in the wider region generally and the potential
for revealing a deep sequence of archaeological layers covering a substantial length of prehistoric time. In that first campaign he excavated
within a 2m × 2m square. It was only when he hit 'bedrock', at the
bottom of the archaeological deposits, some four metres down, that
he stopped digging.

Ovodov had revealed a sequence of archaeological deposits that
reached deep into the Palaeolithic. Initial dating work suggested that
humans were living there as early as 300,000 years ago.

The deep archaeological deposits and the potential for finding more
sites in the region led to more excavations, starting in 1982. This time
they were led by the Russian archaeologist Anatoly Derevianko, a

larger-than-life, friendly and very tough Siberian. I first met Anatoly in Moscow in 2008. He is one of the most famous archaeologists in Russia, leader of Palaeolithic expeditions to Mongolia, Kazakhstan, Montenegro, Vietnam, Uzbekistan and, of course, to the Altai. Anatoly had risen to the very top of the Russian Academy from humble beginnings and become an 'Academician'. The day I met him he brought with him an assistant bearing a stack of books, including one written on his life and describing his achievements and prize-winning exploits, with a bibliography listing hundreds of his publications. On the back was a picture of him with Vladimir Putin and another with Hillary Clinton. This was clearly stratospheric stuff in terms of the world of archaeology. I could not help but be impressed; Anatoly was basically a rock star archaeologist.

A man of huge energy and organization, Anatoly had not only pushed forward with continuing the excavations at Denisova in the wake of Ovodov's initial work, he had also directed a programme of surveying to find other sites in the wider Altai region. He focused on two large river basins, one drained by the Anui river, and the second the Ursul to the south. Over twenty years, his teams worked to identify and explore new sites in these areas. They succeeded in finding ten open-air sites and six cave sites in an area of around 100km^2 in the north-western Altai region. Anatoly built interdisciplinary teams of researchers to work on the vast amounts of newly excavated material and publish the results. The Institute of Archaeology and Ethnography in Novosibirsk at Academgorodok (Academy Town) continued to grow as the pace of discovery quickened.

Academgorodok is a large science city on the outskirts of Novosibirsk. It was founded in 1957 and provided a concentration of scientific institutions comprising up to 65,000 Soviet scientists and, for the majority, was a privileged place to live and work. Although it suffered after Perestroika and the demise of the Soviet Union, it has undergone something of a rebirth since, with much investment. Every time I visit, things seem to have changed, with new buildings shooting up. Usually I prefer to stay at the small museum of the Institute, which has three rather spartan guest rooms, rather than the usually empty Soviet-era Golden Valley hotel, where, when I first came to stay, the 'hot' water took thirty minutes to become even tepid and there was the constant feeling that the room had been bugged by Cold War spies. (That said, even the Golden Valley has improved recently.)

The museum contains some of the treasures that have been excavated by the Novosibirsk teams over the last few decades: the contents of Mongol cemeteries and Bronze Age kurgans or burial mounds, fabulous tattooed Scythian ice mummies complete with preserved skin and hair, lying in long perspex tubes covered with black blankets to keep out the light, and ancient Palaeolithic hand axes and flint blades. Sleeping just down the corridor from some of the spectacularly preserved mummies always gives me a little chill down the spine if I wake in the middle of the night . . . too many horror movies in my teens.

From the mid 1980s, but particularly from the early 1990s, Anatoly and his co-director Michael Shunkov kept Denisova under regular excavation. Initially, they concentrated on the Main Chamber, a

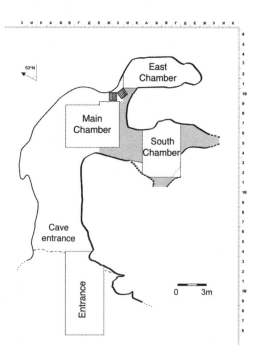

Figure 8 Plan of the different chambers and areas of excavation at Denisova Cave. The grey shaded areas are remaining sections of recent Holocene sediment.

huge arched vault within the marbled limestone cave reaching 10m in height and 11m × 9m across. In 1982–3 the upper layers of the site were excavated, covering the Holocene period – the last 10,000 years. Later, in 1984, and again in 1993–5, 1997 and 2016, the lower deposits began to be explored.

In 1984, towards the base of the excavation in layer 22 of the Main Chamber, a human tooth was found. It was a very worn deciduous molar, well preserved and unlike the tooth of a modern person; in fact, several on the team thought it bore similarities to a Neanderthal tooth. They could not be sure, though, so it was just designated a tooth of an archaic human, which simply means human, but not like us. More than thirty years later this tooth would be analysed using cutting-edge methods for the analysis of ancient DNA and shown to be the tooth of a new species of human: a Denisovan. The tooth became known as 'Denisova 2'.[2] This is the oldest human remain yet found in the Altai region.[3] We know that the tooth came from a girl, based on the genetic analysis, and she was ten to twelve years old at the time she lost her tooth.

The other two chambers in the cave were opened for excavation later. The South Chamber was excavated between 1999 and 2003, and again from 2017, while the East Chamber was worked on from 2004 to the present day. Over the years, many other tantalizing and interesting finds have been made in all areas of the cave. In 1998 an incredible fragment of worked dark-green chloritolite stone, interpreted as being part of a bracelet, was found in the Upper Palaeolithic layers, dating probably to between 35,000 and 45,000 years ago. Usually one would see these types of polished stone objects as being from

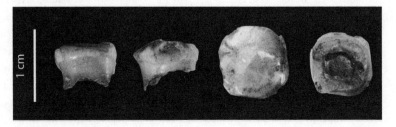

Figure 9 The Denisova 2 tooth, excavated in 1984 in the Main Chamber.

the much later Neolithic (New Stone Age; ~6000–5000 BC),[4] the period where we see the adoption and spread of agriculture. The bracelet, however, was found in much deeper levels than expected if it is from the Neolithic, although it could have moved downwards into deeper Palaeolithic levels over the centuries due to water action or the influence of animals, as sometimes happens in cave environments. In caves like Denisova there is evidence for the presence of animals such as cave bears over much of the last 200,000 years. Sometimes they build nests and dig up the ground beneath them, moving artefacts and archaeological evidence. Anatoly was convinced, however, that this was not an influence in the case of the bracelet fragment. If it is as old as they think, it is a stunning and very unexpected discovery. It might require us to rethink when these types of tools were first used and what purpose they were used for.

One of the interesting aspects of the bone material that has been found at Denisova so far is its fragmentary nature. Around 95 per cent of the bones are so fragmented that they simply cannot be identified to a type of animal, or indeed human. This is probably due to the presence of the main prehistoric predator in the region, the hyena, but bears and wolves also contributed. Hyenas have crushingly powerful jaws and are adept at chomping up bones and then passing them through their guts. In Denisova, we often find bone with gnawing or chewing marks, and many with pitted surfaces derived from the stomach acids of animals. On occasion, hyenas will eat pieces of deer antler, stomach acids etch the tips and once they are passed out and deposited in the archaeological site, they give the impression of looking like humanly fashioned spear points.*

It is astonishing to think that hyenas were once spread widely across Eurasia when we think of them today only as an African animal. The Eurasian hyenas are taxonomically identified as *Crocuta*

---

* In Britain, at the site of Pin Hole in Derbyshire, dozens of these points have been found, all of them derived from hyenas rather than people. Once, on a visit to a site in France, we had to inform a colleague that the precious bone/antler point they'd excavated and analysed was not an artefact at all, but the product of a hyena's stomach. It did not go down well with the master's student, who had produced an entire dissertation on that 'artefact'.

spelaea, or *Crocuta crocuta spelaea*, which roughly translated means 'cave hyena', owing to their propensity to den and have young in caverns. Hyena appear to have comprised two major groups or clades, one in Europe and one in the east, around the Altai.[5] Often, we find evidence for hyena in the same caves that humans were occupying. This almost certainly means that they were competing with each other for access to the cave. I imagine it must have been a terrifying experience for a Stone Age hunter-gatherer to arrive at a cave after a period away and have to enter the darkness to check whether bears or hyenas were living or hibernating there.

There were other carnivores in Eurasia that we would consider African today as well. Evidence for cave lions and leopards is found throughout the continent, as well as for rhinos, albeit woolly ones. These species are no longer present, of course, but how and why they disappeared or contracted in these regions is still being explored. The most common explanations concern climate change and the impact of human hunting. It may have been a combination of both.

The deep sequence at Denisova, comprising twenty-two layers of archaeological deposits, has revealed the changing fortunes of the people who lived there over 300,000 years: what they ate, the types of tools they made, when they were there and what the climate was like. The thousands of animal bones that have been excavated tell us a great deal about what the weather was like locally, because some of them are sensitive to specific types of environment.[6] Some animals, such as red deer, are present in warmer climates only, while others, such as mammoths, woolly rhino and reindeer, were easily able to cope with colder climates. Some of the animals found prefer open steppe landscapes and others more forested areas. The analysis of pollen grains found in the sediment is very informative too. The pollen in the site comes from at least thirty species of tree, as well as an equal number of grasses and shrubs. This allows scientists to reveal the changing nature of local vegetation as it responds to the variations in local climate. And what is certain is that these climate changes have been dramatic over the last 300,000 years, and must have influenced humans and their way of life significantly.

How can we reconstruct the nature of these climatic changes and whether they were violent or more benign? We are fortunate to have several global records of climate change over the period. These come from variations in the ratios of oxygen isotopes measured in deep-sea sediments and in annual layers of ice at the poles. There are two main types or isotopes of oxygen, ^{16}O and ^{18}O, differentiated by two extra neutrons in the nucleus that make ^{18}O the heavier of the two. When seawater evaporates it favours the lighter isotope, so more ^{16}O finds its way into the atmosphere, leaving the seawater progressively enriched in ^{18}O. We call these changes in the relative isotopic proportions 'fractionation'. Tiny shelly animals called foraminifera, or forams, faithfully record the ratios of these two oxygen isotopes during their lifetimes; the warmer their environment, the more their isotope ratio swings to ^{18}O. When they die, they sink to the sea bed and become incorporated in the marine sediment ooze. Over thousands of years this sediment builds up in a layered record. Scientists can core down into these sediments, pull up a core, extract the forams and measure the changes in their isotopes in increasing depths. We can also date them, thereby creating a long marine palaeo-temperature record.

When evaporated rainwater cools as it moves into the higher latitudes of the Earth it condenses. Just as water containing ^{16}O is more likely to evaporate from the ocean so the ^{18}O in water is more likely to condense to form rain or snow. This difference is strongest at lower temperatures thus the oxygen isotopic composition of snow and ice

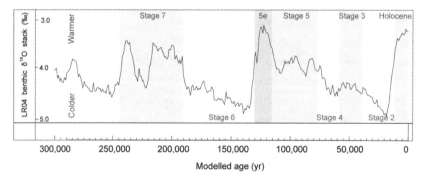

Figure 10 The record of climate change over the last 300,000 years based on oxygen isotope variations in marine foraminifera. The higher the peak, the warmer the temperature.

in the polar regions provide a key temperature record at the poles, with more oxygen 18 in ice during times of colder climates compared with warmer periods. As in the marine environment, scientists have bored down through the annual layers of packed ice, more than 3,000m in the case of some of the recent deep cores, right to the bedrock of Greenland. By measuring the oxygen isotope ratios in trapped CO_2 in these layers of ice they have created a record of temperature changes spanning 125,000 years. In the Antarctic the record is even longer – at least 800,000 years.

These various sources of evidence contribute to understanding how the climate and environment around Denisova varied over long periods. Sometimes conditions were similar to the present day or even warmer, and broad-leafed forests spread across the Altai (such as in Stage 5e in Figure 10). When the first Denisovans came to the cave they enjoyed these very conditions, during Stage 7. At other times, however, the weather was icy and glacial conditions were present. Forests melted away and were replaced by plants typical of the cold of the taiga or steppe environments, with animals such as lemming and reindeer present. Sometimes the climatic changes came with terrifying speed. Evidence from Greenland suggests that switches in the climate system occur roughly on a millennial timescale with dramatic increases in temperature occurring in as little as two or three years.

We can determine what species the people at Denisova preferred to hunt by carefully analysing evidence for cutting and butchery marks on the bones excavated in the site. The location of Denisova is right at the point where a narrow gorge begins. As animals move across a landscape following new pastures or migrating, these pinch-points can be attractive locations for human hunters, so Denisova was prime territory for people. It was also one of the rare caves in the region, so was in many respects an optimal place to live. A wide range of bone tools and artefacts have been found that suggests humans processed wood and prepared skins and plant fibres in Palaeolithic times. Later, in the upper parts of the Palaeolithic levels, after 45,000 years ago, we see tools and artefacts appear that have a decorative purpose. Small pierced beads made of ostrich eggshell are periodically identified, for example, that may have been used to decorate clothing or been laced and worn around the neck. Like the hyena, the ostrich was much more widely dispersed in the Pleistocene* or Ice Age. Research has shown that they probably disappeared from Eurasia as recently as 10–12,000 years ago. Early humans spotted their eggs and used them to fashion jewellery. We have found ostrich eggshells in Siberia, Mongolia and Central Asia from as early as 40,000 years ago.

The Denisova team has also found the pierced teeth of fox and reindeer, bison and deer, again used for decoration. There is a double-drilled ring fragment made of ivory from a mammoth tusk, tubular bird bones with regular incised markings (for what purpose we simply do not know), marble rings and bone needles – ideal for threading and perhaps creating fur clothes, leggings and boots. I was in the Main Chamber in 2016 when one of the excavation team found a complete bone needle, 7.5cm long. It's incredible to hold a freshly

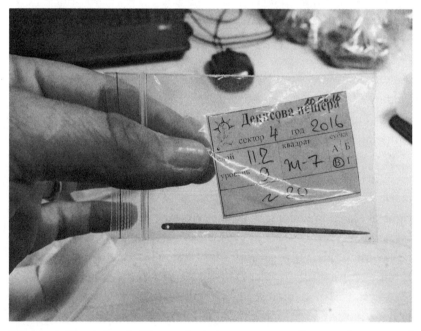

Figure 11 The author holding a plastic bag containing the recently excavated remains of a perfectly preserved bone needle from Denisova Cave.

---

* The Pleistocene refers to the epoch covering the last 2.6 million years up until 11,700 years ago.

excavated object, made probably 35–40,000 years ago, which is so delicate and beautiful. I often imagine the time that has elapsed since that object was held by the person who last used it in the remote past, in terms of human generations, as a way of thinking about the span of time that separates us. If we think of a human generation as around twenty-two or twenty-three years, then more than 1,700 generations have lived and died since those times. I think about the people who made these objects, their families, children and grandparents, and how their daily lives were structured.

What did they think? What hopes and dreams did they have? What had they to do to survive in the dramatic climatic and environmental changes that have characterized much of the last 100,000 years? More recently I have also started to wonder which kind of humans might have made these items. Usually, we see items like this as being made by our own species, *Homo sapiens*, but as we shall see in this book, this is still tantalizingly unclear.

In July 2008, in the East Chamber at Denisova, a tiny bone was found by one of the diggers in the cave. Anatoly Derevianko and Michael Shunkov thought it might be a tiny vestige of an anatomically modern human. The bone is a finger phalanx; the final bone of the smallest, 'pinky' finger. It's probably from a person around thirteen years of age. We now call this bone 'Denisova 3' or the 'Altai Denisovan'.

The following year, a small sub-section of this bone would make its way to Germany, to the Max Planck Institute in Leipzig, where its DNA would be extracted. Ancient genomics has been revolutionized over the last twenty years. We will explore what this means for archaeology and tiny human remains like Denisova 3 in Chapter 6. First, I want to introduce the key aspects of the genetic revolution, how it works and what it can tell us about the deeper archaeological past.

5. The Genetic Revolution

Ancient DNA or genomics has exploded in the field of human evolution. It has revolutionized archaeology, providing insights into the past that have yielded seismic shifts in our understanding. I've been fortunate to be an eyewitness as the science of ancient DNA began to be improved and then more and more widely applied. This coincided with my arrival in Oxford, where I met and worked with team members in the university's Henry Wellcome Ancient Biomolecules Centre. Through and with them, I was able to follow closely the development of the science. In this chapter I am going to introduce some quite complex concepts, which might be a bit heavy going at times, but it is important to understand at least the basics of genomics, and it will also come in handy in reading the rest of the book. I hope very much that, by the end of this chapter, if I talk about how many SNPs were identified in an ancient genome, you'll know immediately what I mean.

In the late 1980s I remember first hearing about the potential for extracting ancient DNA for analysis from human bones. Initially the results seemed incredible, and in some cases almost too good to be true.[1] Gradually, the work of several scientists showed that, sadly, some were. Often, geneticists who had worked on bones had unwittingly contaminated them with modern human DNA, and therefore obtained essentially meaningless data. In some instances, ancient DNA sequences were shown to come from bacteria on the floor of the laboratory rather than having an ancient origin. There were calls for a radical improvement in laboratory protocols and duplication to avoid the publication of problematic results.[2] Several colleagues working in the field told me that they thought it would never be possible to extract verifiably ancient DNA from humans due to these contamination issues. Even touching a bone briefly with one's fingers, it was said, would introduce contamination that penetrated to the very centre. My late colleague Roger Jacobi sometimes used to discreetly lick bones to check whether they had been conserved or glued before we sampled them for radiocarbon dating. Interestingly, in the 1850s bone licking was commonly applied because it was thought to be a tried and trusted method of dating them. An ancient bone would stick to the tongue, unlike a modern equivalent, so many researchers used to habitually lick bones to assess antiquity. The most famous example of this was the attempt to date the first Neanderthal, the Feldhofer specimen in Germany, that we met in Chapter 3.[3] As well as licking, imagine the many people in excavations and museums who had touched bones directly; surely bones would be contaminated for evermore.[4] The problem researchers had to contend with was how to authenticate the DNA extracted, to make certain it was genuine and not from a modern or contaminating source.

Thankfully, over the last few years, things have changed dramatically and become much less bleak. Scientists have worked out ingenious ways of eliminating contaminants and the DNA fragments derived from other organisms such as microbes and bacteria, and selecting only the endogenous, or native, DNA for analysis. Great care in ensuring a clean environment for extracting the DNA is paramount. Laboratories are built with 'positive pressure' in them so that air does not rush in from the outside when a door is opened. Scientists also go to great lengths to wear protective clothing in the laboratory to ensure they do not contaminate bone and teeth samples with their own DNA. Visiting a clean lab requires using overshoes or slippers that are allowed only in the laboratory space, as well as donning extensive PPE (personal protective equipment). Our bodies liberally shed DNA, so it is crucial to seal them off from the vital samples. The almost pathological approach of several laboratories – most importantly for our story, the Max Planck Institute facilities in Leipzig – in overcoming the pervading issue of contamination is one of the key reasons why we can now extract demonstrably authentic ancient human DNA.

Geneticists have also worked out how to extract DNA that has evidence of being 'damaged' or chemically altered over time, which means it is therefore demonstrably ancient rather than derived from modern, contaminating DNA.

To understand how these decontamination and extraction approaches work, we must understand the fundamentals of DNA itself. The DNA molecule is like a twisted ladder. The uprights of the ladder are formed of deoxyribose and phosphate molecules that alternate with one another. The rungs are formed by the so-called DNA bases, or nucleotides: adenine (A), which is always paired with thymine (T), and guanine (G), always paired with cytosine (C). Over time some parts of the rungs of the ladder undergo a chemical modification. The DNA base cytosine, for example, can be converted chemically to uracil, which, when the DNA replicates, goes on to pair with adenine instead of guanine. Importantly, the frequency of these uracils in the DNA is highly correlated with time: the more ancient the bone, the more uracil substitutions it has. When the DNA sequences are read in the laboratory, the enzyme that is used to translate the sequence will read a T across from that A, instead of the G that cytosine is always paired with.[5] In the sequence, therefore, C has magically changed chemically to a T. These C-to-T substitutions result in a higher proportion of Ts at the ends of each ladder of DNA. It sounds bad but, crucially, this higher proportion makes it very likely that these strands of DNA are in fact ancient and therefore uncontaminated by modern DNA. Geneticists can physically separate the DNA molecules in the laboratory that have high uracil-influenced counts, knowing that these are the ancient molecules rather than the contaminating ones.[6]

In addition, it is much more likely that ancient DNA will consist of short sequences of base pairs rather than very long ones. This is because the longer chains of bases are more likely to derive from modern DNA and therefore be a contaminant, while the shorter sequences are short owing to age and degradation. Recognizing this has allowed researchers to focus on eliminating longer contaminating DNA sequences prior to analysis.

There are other chemical tricks to remove contaminants and raise original or endogenous DNA yields too, including the use of chemical cleaning treatments such as bleach on the bones prior to DNA extraction.[7] These, and other changes, have meant that since about 2003 we have been able to extract demonstrably ancient DNA sequences from human bones (I should say that my formerly sceptical colleagues are thrilled to have been proven wrong).

The second major development in ancient genomics has been the development of powerful sequencing instruments which make it possible to sequence the genetic code reliably on an almost industrial basis.

In the 1990s scientists primarily used a method called PCR, which stands for polymerase chain reaction. This method is still used today, in fact the Covid-19 pandemic testing done in 2020 largely used this method to detect RNA associated with the virus. It is a technique that enables small fragments of DNA to be copied using a type of enzyme called polymerase, increasing or amplifying small pieces of DNA to allow them to be analysed more easily. This technique was revolutionary, and indeed the developers, Kary Mullis and Michael Smith, were awarded the Nobel Prize in Chemistry in 1993 for their work developing it.[8] It enables vast amounts of DNA to replicated, or amplified, over and over again. Polymerases use the old DNA as a template to synthesize a new copy. The copy is ingeniously built from new nucleotides (or bases) that are knitted together by the polymerase starting from what are called primers. These are short single strands of DNA (around twenty bases) that are bound to the end of one part of the ancient DNA fragment. Rather like a zipper, the polymerase's job is to take free single deoxynucleotide bases that are added in the laboratory in test tubes and join them chemically to the fragments of DNA, in sequence, one by one, thereby amplifying the amount of DNA one has.

So much for amplification; how does one extract the actual known sequence of nucleotides, the A, T, C, G, or adenine–thymine and cytosine–guanine, the exquisite code of DNA that forms the double-helix – otherwise known as the tree of life – *and* make sense of it? To understand this, we first have to understand sequencing, and for this we have to go back to 1977 and the development of the first gene-sequencing technology by Nobel Laureate Frederick Sanger.*

---

* Sanger, who declined a knighthood, won two Nobel prizes in Chemistry, one of which was for what would become known as 'Sanger sequencing'.

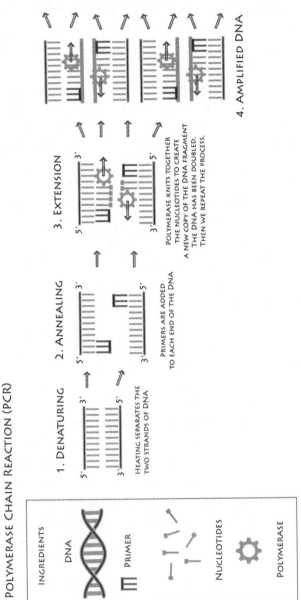

Figure 12 Schematic of the polymerase chain reaction.

The way in which the precise sequence of letters in the DNA is built relies upon a clever method of stopping the PCR at the exact place that corresponds to the base you want to read. The reaction is stopped by adding a different type of deoxynucleotide that is not able to form a chemical bond with the next base in the sequence. These are so-called dideoxynucleotides or ddNTPs.

The fragments of amplified DNA are divided equally into four test tubes, each of which contains deoxynucleotides with either adenine, thymine, guanine or cytosine. Each test tube then receives a single type of the respective ddNTPs, or chain-terminating nucleotides. So the first test tube only has an adenine-labelled ddNTP, the second a cytosine-labelled ddNTP, and so on. In the first test tube, therefore, the replication of DNA continues until the reaction stops at the point where a specific base-labelled ddNTP is added to the sequence of that DNA fragment. You therefore end up with four test tubes containing DNA fragments of variable length which terminate in every place you have an adenine, thymine, guanine or cytosine base in the sequence.

To work out in which position in the overall DNA sequence these bases now sit, it is necessary to quantify the sizes of each fragment, and for this we use the next part of the Sanger process, which is electrophoresis. This simply describes the movement of different particles within a liquid under a constant electrical charge. In this case the liquid is a gel called polyacrylamide. The contents of each of the four test tubes is added to the gel in four different lanes, and then a charge is applied. Slowly the DNA moves down each lane from the negative to the positive pole of the charged field. The smaller and lighter the fragments are, the further they travel, but when they stop they create visible bands. In the first lane of the gel are the fragments that stop with the As, the second lane, only the Cs, then the Gs and finally, in the fourth lane, the Ts. Once completed, the sequence of bases can be read from bottom to top, in order of mass, and then put into the correct order based on their known paired base. From this one ends up with the DNA sequence – AGTTCAGCATAGA etc. This method was used to generate the human genome in the ambitious Human Genome Project that took a decade and cost $3 billion.

SANGER SEQUENCING

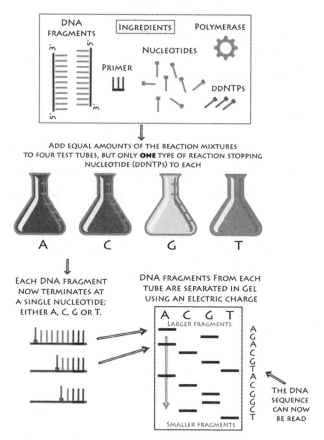

Figure 13 Schematic showing Sanger sequencing.

While being a brilliant development, PCR does have some draw-backs, particularly for ancient DNA. As well as amplifying the ancient DNA, it also amplifies the contaminating DNA; rather like the manure you add to the garden improves both your roses and your weeds at the same time. This was the major limitation for analysing the DNA segments extracted from ancient bone and teeth in those early years.

Subsequent developments in sequencing technology led to signifi-cant changes and have been the game changer in the ancient DNA world. The US biotech company 454 Life Sciences was in the vanguard of this revolution. It is said that CEO Jonathan Rothberg came up with the idea for improving DNA sequencing when one of his children fell ill with breathing difficulties. Rothberg was frustrated that doctors could not diagnose quickly enough whether his child's illness was an inherited condition or not. A full genome would potentially reveal the reason. While reading a magazine in the hospital Rothberg noticed a picture of Intel's new Pentium microprocessor on the cover. He decided that speeding up sequencing by working on numerous DNA snippets in parallel alongside greater processing power was the way to improve genetic analysis, so he founded 454 Life Sciences, whose aim was to improve sequencing speed. Later, 454 would be a key partner in the Neanderthal Genome Project, the programme designed to sequence the Neanderthal nuclear genome.[9] Rothberg made a huge amount of money from this developmental work, and other scientific adventures. Famously, he spent some of it on recreating Stonehenge at his house on Long Island. Named the 'Circle of Life', it is made of 700 tons of imported Norwegian granite.

The newer sequencing approaches of companies like 454 markedly sped up the process, making it around a hundred times faster. There are several crucial differences from the Sanger approach. The first is improved chemistry. So-called 'adapters' are added to both ends of short fragments of DNA from the sample. These allow the sequencing machine to recognize the beginning and end of each DNA fragment. The sequencing is performed on a small plate, rather like a microscope glass slide, which has hundreds of tiny wells in it within which are small resin beads with a complementary sequence to the adapters, enabling them to stick to the adapter at one end of each fragment of DNA and hold them in place. Afterwards, waves of compounds containing the nucleotide bases are added sequentially; first As, then Cs, then Gs, then Ts, and as with Sanger sequencing the bases are knitted one by one to the DNA sequence using the polymerase. A key difference though is that the electrophoresis step is eliminated through a method called pyrosequencing. This makes it possible to detect the bases far more efficiently. When a nucleotide is successfully added, pyrophosphate is released and chemically converted into a small flash of light. A very precise camera detects each flash on the plate and, by measuring the intensity of the flash, records which kind of base has been joined to the sequence. In this way the sequencing process is rapid and continuous.[10]

Using these approaches, tens of millions of bases could be read per day. This method, and subsequent modifications to it, is known as next-generation sequencing, or NGS, and it has led to a huge surge in the application of ancient DNA methods to the archaeological past. New and improved sequencing platforms, made by companies like Illumina, offer even faster and more automated options. Instead of using expensive enzymes, these sequencers now record the base added to the sequence using a dye corresponding to that base, and whose colour can be rapidly read by high-resolution digital cameras. Instead of a genome taking ten years to sequence, like the original human genome, this can now be done in a single sequencing run and take only a day or two. Tens of thousands of genomes can therefore be determined per year. Ancient DNA is much trickier, however, because it is often so fragmented and therefore the pieces of DNA are shorter and more time is needed to sequence.

Now we know about the sequencing technology, we need to understand the type of information that DNA in ancient bones can provide.

There are two main archives of genetic information in human bones. First, mitochondrial DNA (mtDNA). These are small circular-shaped chromosomes found in the mitochondria; the energy-generating parts of individual cells. The second is the nuclear genome, which is found only in the cell nucleus. This is diploid or double stranded DNA, which means we receive half of the DNA from our mother and half from our father. This nuclear genome is the product of generation after generation of mixing, a process called 'recombination'. In our nuclear DNA, therefore, we hold a record not just of us, but of huge portions of our own genetic history, as the playing cards of genetic information are split and re-shuffled generation after generation on their way to us. A single nuclear genome is therefore incredibly informative in terms of deeper population history. Mitochondrial DNA, on the other hand, comes to us from our mother, who gets it from her mother, who gets it from her mother, and so on back. The much larger size of the nuclear genome means that it is substantially more informative than the mito-chondrial DNA. While the mitochondrial genome, or mitogenome, comprises around 16,500 base pairs of information, the nuclear genome is almost 200,000 times bigger, a whopping 3.2 billion base pairs.

It's difficult to get an idea of the size of this latter figure, so to help, let's imagine that we decided to write a book containing all of the base pairs in one person's sequence of A, C, G, Ts, in order, from beginning to end. If we assume that we could fit 2,000 letters on a single page, we would need 1.6 million pages of text, and if each book were 500 pages long, we would need to have 3,200 books to produce the genome in its entirety. A mitochondrial genome of data, by comparison, would fit on just over eight pages of text. It ought to be said, however, that vast amounts of the nuclear genome appear to be largely uninformative, sometimes called 'junk' DNA.★ Much of it is shared across many dif-ferent species, so geneticists often focus on the variable regions which are called SNPs (single nucleotide polymorphisms – pronounced 'snips'). These are much more informative owing to their variability between different people and groups. Often you will hear that geneticists have sequenced 250,000 SNPs in order to compare different human genomes, and this refers to these variable sections of genetic code. Commercial DNA sequencing companies will sequence several hun-dred thousand SNPs from your genome for you, if you pay them.

A final word is needed on the analysis of the generated data from all this sequencing. The sequences of letters in the DNA can be com-pared with other sequences in order to probe differences and their significance. This is called sequence alignment. The letters in the sequenced code are laid in rows. Other sequences can then be added in rows below for comparison. When properly aligned, differences between the sequences at specific points in the lines of letters become visible. These can reflect point mutations (a place where a base has changed due to a mutation), or be caused by insertions or deletions (indels) of DNA code, usually ranging in size between 1 and 10,000 bases long. These can be derived from admixture or interbreeding

---

★ I hope no readers will infer from this analogy that there must be vast amounts of junk text in this book.

with another, related species, for example. Powerful mathematical algorithms are required to help probe the high-coverage sequences of DNA that are increasingly extracted and compare them with other sequences to derive the significance of the differences.

Since 2013 the number of working human genomes from the archaeological record has multiplied exponentially thanks to the NGS revolution (see Figure 14). This science is making huge contributions to archaeology because it is enabling sequencing to become more and more routine by the month.

I was a co-author on the first so-called 'complete human genome' paper in 2010,[11] of the DNA sequence of an ancient Greenlander who was nicknamed 'Inuk'. This work was led by Eske Willerslev in Copenhagen. Such was the impact of this paper that it was published in *Nature*, the prestigious science journal, and featured on the cover. Eight years later I was a co-author on another paper in *Nature* about the genomics of the Beaker people of Bronze Age Europe, which

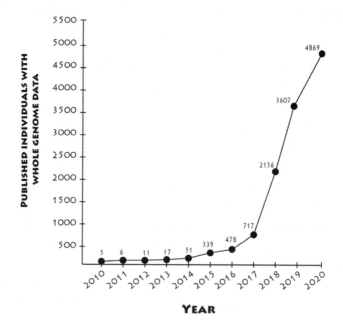

Figure 14 The rise in the number of whole ancient genomes published over the last decade, 2010–20 (note that 2020 figures are only for the first 6 months of the year) (data courtesy of David Reich, pers. comm.).

featured more than 400 ancient human genomes.[12] No one could have predicted that in such a short time we would be dealing with such massive numbers of genetic datasets, but it has come to pass seemingly in the blink of an eye.

The impact of genetics upon the findings in Denisova Cave has been particularly profound. It has turned it from another Palaeolithic cave somewhere in Siberia into one of the most famous and important archaeological sites on the planet, and one that all prehistorians are talking about. The advances in technology that we have described led directly, and in many respects inevitably, to the discovery of a new human species, as we shall see in the next chapter. Archaeologists had been waiting for these developments in DNA analysis for decades; they were about to reap the rewards of that patience.

6.   A New Species of Human

On Friday 4 December 2009, at the world-leading genetics laboratory of the Max Planck Institute for Evolutionary Anthropology in Leipzig, something extraordinary happened.

A young post-doctoral scientist called Johannes Krause and PhD student Qiaomei Fu were sequencing DNA from a tiny (7mm × 5mm) bone sample that had come from the fifth digit of a human hand. The bone was part of the same specimen that had been spotted in 2008 during the excavation of sediments from layer 11.2 in the East Chamber of Denisova Cave, mentioned briefly at the end of Chapter 4. Owing to its potential significance, it had been hand delivered by a Russian archaeologist to Johannes's doctoral supervisor, Svante Pääbo, Director of the Department of Evolutionary Genetics at the Max Planek, leader of the Neanderthal Genome Project and a pioneer of the field of ancient DNA. After extracting and sequencing the mitochondrial DNA, Krause aligned the sequence of the Denisova mtDNA against six of the Neanderthal mtDNAs that Pääbo's group had analysed previously, as well as hundreds of modern human mtDNAs that had been studied of people alive today.

That's when he noticed something unusual.

While Neanderthals differ from modern humans at an average of 202 base positions along the 16,500 base pairs of the mitochondrial genome, the sample from Denisova cave differed at 385 positions, almost twice as many. This could mean only one thing: Krause must be dealing with something completely different. A new type of so-called 'archaic' hominin, not a Neanderthal and not a modern human like us; something never before seen. It was, he recalls, a dramatic and unforgettable moment. He immediately rang Pääbo, who was away from the laboratory at a conference, to give him the news. Prior to delivering his bombshell, Krause told Pääbo that he had better sit down.[1]

It is rare in science that one finds a group of never-before-seen humans, but no one had ever discovered such a thing in a scientific laboratory. Usually these kinds of moments in palaeoanthropology are reserved for fieldwork, when a trowel reveals the mandible of an ancient hominin or archaeologists surveying a site in the African Rift Valley come upon a piece of human fossil or a partial skeleton. This was a real first. Krause remembers the occasion as 'scientifically the most exciting day of my life'. Pääbo recalls being dumbstruck when told, but extremely excited. He rushed back to Leipzig to discuss the results with Krause and the rest of his team.

Usually, it is customary in scientific taxonomy, when one names a new species, to have what we call a holotype. This describes the type or example specimen, the one that everyone refers to henceforth when referring to the species, like the Feldhofer Neanderthal. Often these holotypes are described and reported with great fanfare and excitement in the top academic journals. In this case there was no holotype, only the tiny remnant of the finger phalanx designated 'Denisova 3'. What was one to do?

Pääbo and Anatoly Derevianko, along with the wider team, initially decided to go with naming the new species *Homo altaiensis*, and included this in the draft paper that they submitted to *Nature*. Some reviewers of the paper, however, had doubts about whether it was wise to use genetic data as the basis on which to name a new species. What if it turned out later that they had sequenced an already taxonomically defined species, such as *Homo erectus*? They therefore decided to withdraw this until such time as the picture had become clearer and simply refer to it as an 'unknown hominin'.[2] Pääbo and Krause were at pains to state that any definitive conclusions regarding the species status of the Denisovan bone would have to wait until the far more informative nuclear genome was probed. And this was a big 'if', because of the much more complex technical challenges associated with it.

The paper was published in *Nature* on 24 March 2010.[3] It was a sensational revelation, of that there is no doubt. I remember reading it several times, quite stunned that somehow, somewhere, there was once another human group out there in Eurasia. How could archaeologists have missed this? Could it really be true? Surely we would have found the bones of such a human cousin before now. Svante Pääbo's words at the time mirrored my thoughts entirely: 'I almost could not believe it. It sounded too fantastic to be true.'

I emailed my father the day the paper came out to tell him the news. He said it was 'stunning, [I] will incorporate in my stage 1 lectures'. That meant it must be big!

As the paper was submitted and then published, Pääbo's research team began referring to the specimen as 'X-Woman'. They decided they must begin the job of painstakingly attempting to sequence nuclear DNA from the tiny remaining bone in order to place it more reliably amongst the then-known assemblage of other prehistoric hominins.

Later in the year, on 24 August 2010, an email arrived from Magdalena Skipper, an editor at *Nature*, asking me if I would help review a newly submitted paper. Usually, a new *Nature* paper contains something exciting and novel, so of course I was immediately interested in what this one concerned. The first author on the paper was someone named David Reich from Harvard University, a name new to me, and it was about Denisova. The abstract pasted into the body of the email below read:

Using DNA extracted from a finger bone found in Denisova Cave in southern Siberia, we have sequenced the genome of an archaic hominin to about 1.9-fold genomic coverage. This individual is from a population that shares a common origin with Neandertals but has a distinct history. This population contributed genes to Melanesians and may itself have received genes from another more archaic hominin group . . . We designate the group of hominins to which these individuals belonged 'Denisovans' and suggest that they may have been widespread in Asia during the late Pleistocene.[4]

My jaw dropped. They'd got *nuclear* DNA out of the specimen? In just a few months? And they had found that Melanesians had a proportion of their DNA derived from these 'Denisovans'? Again, it was hard to keep up with the implications and the results. If the initial 'X-Woman' paper was a bombshell, this paper was a series of firecrackers going off in all directions. The team not only had sequenced nuclear DNA but also had covered virtually every base in the nuclear genome an average of 1.9 times. And in the nuclear genome there are more than 3 billion base pairs to map, of course. We often denote this coverage with an 'x', to show the number of times on average each base position has been read, i.e. 1.9x coverage in this case. The paper went on to say that it was the high level of endogenous or original DNA that could be recovered from this tiny bone that made this possible. Even so, this was a technological breakthrough in terms of ancient DNA, and from a specimen tens of thousands of years old.

Denisova 3 yielded 70 per cent endogenous DNA. This figure is astoundingly high in palaeo-genomic work. It is more common to see endogenous DNA in the order of under 1 per cent, perhaps up to 5 per cent in exceptional cases, because most of the DNA extracted from ancient bones is nearly always bacterial in origin. But 70 per cent? It was almost modern in terms of its preservation. Incredibly, the team had done their work on a bone sample weighing just 40mg; that's about one and a half rice grains. It's an extraordinarily small amount. And from this, they sequenced the nuclear genome.

The 2010 Reich paper is a landmark publication, from a technical and scientific basis as well as from a palaeo-anthropological perspective, and one that palaeoanthropologists keep dipping back into because there is simply so much information in it that is of deep interest. Later, in 2012, a new method was applied to the DNA sample library of Denisova 3 that saw even more genetic data generated. A much higher-coverage genome was extracted using a novel approach in which single strands of DNA were analysed, substantially increasing its yield. This resulted in a 31x high-coverage genome.[5]

There are often interesting stories behind the articles soberly published in scientific journals, and the story of Denisova 3, the now famous pinky bone, is a nice example. When Pääbo's team realized that they had something entirely new on their hands they flew immediately to Novosibirsk to meet with Anatoly Derevianko and the Denisova team to discuss the results and what to do next. When they asked for some extra material from the finger bone they were told by Anatoly that he had sent the rest of the bone, the bulk of it, to another genetics laboratory, in Berkeley, California, run by a scientist called Ed Rubin. Rubin had initially been part of the research team working on the Neanderthal Genome Project, the huge scientific effort led by Pääbo to sequence the genome of the Neanderthals.[6] So the Berkeley team had the other part of this crucially important specimen. Pääbo was disappointed and somewhat surprised. It was then that he knew they might be in a race to publish first, because if the Berkeley group had also sequenced the bone then they were probably aware of its huge significance. There was, then, a real issue of timing.

In science, first is everything; second is first loser. What Pääbo and the team did not know, however, was that Rubin and his group had done no work whatsoever on the Altai bone; it had been left in the lab for another time. Pääbo rapidly published his team's mitochondrial DNA paper and then turned to focus all of their efforts on the nuclear genome.

Meanwhile, Rubin had sent the bone sample to one of his palaeogenetics colleagues, Eva-Maria Geigl, at the Institut Jacques Monod in Paris, to be worked on. Following the publication of the mtDNA paper by Krause and Pääbo, she planned to sequence the nuclear genome. Her efforts came to nothing, however, since she could not extract adequate amounts of nuclear DNA. After the Reich genome paper emerged, Rubin asked for the bone back and it was returned to Berkeley in 2011. This is where the trail runs dry. The bone appears to have been either lost or misplaced, and to this day no one has been able to trace the precious sample's whereabouts or discover what happened to it.

Fortunately, before she sent the bone back to Rubin, Geigl had taken high-resolution photographs and further samples of the bone for future work.[7] In 2019, this analysis was published after encouragement from Pääbo. The mtDNA genome Geigl's team had obtained was identical to the sequence published by Krause and others in 2010, confirming that the material she had been working on was from the same piece of bone. The high-resolution photographs enabled the team to reconstruct the bone and to knit it back together 'virtually' alongside the smaller proximal part previously sampled. This re-united bone scan enabled the researchers to work out a much more precise estimate for the age of the Denisovan girl. As we grow from a baby into an adolescent, parts of our bones undergo significant changes. A bone like a tibia is made up of three distinct parts: the diaphysis or shaft is the major component, then there is the meta-physis, which is the flared end of the shaft, and the epiphysis, or end cap of the bone where it meets the femur. At birth, the epiphysis floats in a disconnect from the main shaft of the bone. As new bone is formed and the child's legs grow the epiphysis is only connected by a thin cartilaginous strip. Eventually, though, this hardens or ossifies and in time the tibia becomes one complete bone. This pro-cess is called epiphyseal fusion. By analysing bones from people whose age at death is known, physical anthropologists have worked out the age at which this epiphyseal fusion occurs in different bones, from fingers, to ribs, to femora and more. They can use this to assign ages to bone fragments like Denisova 3. So, by virtually refitting these two pieces back together, and carefully observing the degree of fusion, the researchers could work out how old Denisova 3 was when she died. The epiphysis was fusing at the time of death, a pro-cess that takes two to four months to complete, and so the pinky bone was very close to its final mature state, the state one would observe in a young adolescent. Because the nuclear DNA had shown the bone was from a female, it was possible to compare the bone's size to living female humans and determine that the pinky bone was from a young woman aged 13.5 years. Not only that, but analysis of the bone's tuberosity and shaft curvature indicated that it almost certainly came from the right hand and was indeed the final phalanx of the little finger, or 'pinky' bone. As we will see in Chapter 9, we have also worked out that the bone dates from between 52,000 and 76,000 years ago.

It never ceases to amaze me how much information can be gleaned from tiny archaeological remains: this bone in total is only 2cm long and yet we can reconstruct the age of the person at death, the date range when this happened, the gender, and which hand the bone came from. Of course, we also have Denisova 3's complete nuclear genome, which contains a wealth of other illuminating information, from the relatively mundane (she probably had dark brown hair, brown eyes, along with dark skin – and no freckles!) to aspects of her population history, health and evidence for disease.[8]

So why was the DNA so well preserved in Denisova 3? It is difficult to say, but there are several possibilities. First, the bone itself is from the tip of the fifth finger, so one possibility is that because the extremities are the first to dry and desiccate, they are the least likely to be attacked by flesh-eating bacteria or broken up by endogenous enzymes. There isn't much flesh on a small pinky bone at the best of times, so the DNA may have been well preserved for this reason. Second, the cave itself and the conditions inside may have played a role. If you go inside Denisova Cave today, take a coat, it's cold and the temperature is remarkably stable. The excavation team wear warm overalls and boots, which keep out the chill and give the appearance of astronauts about to enter a rocket. The average ambient temperature fluctuates a few degrees above zero. This favours DNA preservation. DNA is like ice cream, my colleague Tom Gilbert is fond of saying; in cold climates ice cream doesn't melt as fast as it does in the tropics. These two points don't explain everything, though, because many bones in Denisova don't have any DNA left; it's all degraded away. Caves may have localized preservation conditions that vary from one spot to another. Bone is a complex material, composed of different biomolecules, primarily a mixture of a mineral cement called hydroxyapatite and a protein component, dominated by collagen. The degree of acidity, or pH level, the presence of water, microbes and bacteria, and temperature, all variously combine either to favour the preservation of DNA and collagen or to hinder it. In some parts of the cave it appears that organic matter in bone is preserved for science while in others it is not. Serendipitous is the word; in the case of Denisova 3, the palaeoanthropological world seems to have struck very lucky.

The Denisova 3 nuclear genome was able to tell us a huge amount about modern human populations and our deeper history because of the detailed high coverage of the genome. The wider evolutionary relationship between Denisovans and other hominins could be revealed by comparing it with the human reference genome along a line going back millions of years to the human/chimpanzee common ancestor, to work out how far along this line Denisovans had diverged away from us. Reich's team calculated that this was 11.7 per cent of the way back. The same analysis for Neanderthals showed 12.2 per cent, a quite similar level of divergence, suggesting that Denisovans and Neanderthals are likely to be sister groups. Later, it was estimated that Denisovans and Neanderthals split from one another around 420,000 years ago.[9]

As we saw in Chapter 5, because the numbers in the nuclear genome are so huge, a great deal of computational analysis is required in order to probe it extensively. This field is called 'bioinformatics' and it is populated by mathematicians and population geneticists who are adept at working with these big numbers and building algorithms that enable the patterns and differences within the long sequences of bases to be recognized. An array of new statistical approaches is now being applied to help tease out these patterns. One of these is a statistical test called the D Statistic.[10] It was developed by Nick Patterson, one of David Reich's laboratory colleagues, a person whose background made him ideally suited to the world of pattern recognition and delineating underlying patterns in large data sets.

As a young undergraduate in the University of Cambridge in the 1960s, half of Nick's professors had come from Bletchley Park and the code-breaking group that helped the Allies win the Second World War. Initially, he entered that world himself and became a cryptographer in Britain's Government Communications Headquarters (GCHQ) in the early 1970s. Later, he worked in the top-secret Center for Communications Research in the US, until the Cold War ended. He then moved into the world of hedge funds and stock market forecasting, again using mathematical methods to predict where best to invest.* In the early 2000s, fortunately for the world of human

---

* He was apparently successful: in eight years the assets in the company he worked for went from $200 million to $4 billion.

evolution, he left and went to Massachusetts,* where he began working with David Reich on the underlying patterns of human genetics
using mathematical approaches. This time he wanted to work out the
complex story of human populations and their ancestry. His D Statistics approach is now widely used to test the amount of genetic
relatedness between populations (we call this 'admixture') and to
give quantitative estimates of differences and similarities.

Patterson and Reich took the Denisova 3 genome and compared it
with the genomes of 938 living humans from fifty-three populations
all around the world that had genotyped data of around 642,690
SNPs – the sections of human DNA that provide the greatest level of
variation in the nuclear genome.[11] They wanted to see which modern
populations, if any, were most closely related to the Denisovan genome. They identified three principal clusters in the genome
comparisons. The first group comprised sub-Saharan African populations, of which there were seven. The second comprised forty-four
non-African populations and one from North Africa. The third
group included Papuan and Bougainville populations from Melanesia. They found that this last group shared most alleles with the
Denisovan genome, while the second group contained no Denisovan
DNA at all. When the fraction of derived alleles coming from Neanderthals or Denisovans was tested they found that around 2.6 per cent
of the genomes of the non-African populations came from Neanderthals. This had been known, but less precisely, since the results
of the Neanderthal Genome Project had been published back in
May 2010. Reich and Patterson's analysis now showed that some
4.8 per cent of the Melanesian genome derived from Denisovans.
Together that meant that around 7.5 per cent of the genomes of the
Melanesian people in all probability derive from population admixture with these now disappeared human groups. The most likely
explanation for this was gene flow from Denisovans into the ancestors of Papuans and Melanesians at some point in their evolution,

---

* Initially to the Whitehead Institute for Biomedical Research, part of the Massachusetts Institute of Technology (MIT), whose Center for Genome Research
would eventually become the Broad Institute of Harvard and MIT.

and that this happened as our ancestors expanded outwards from Africa and moved into eastern Eurasia after around 50–54,000 years ago.

Let's pause for a second and let that sink in, because it is quite a stunning revelation. Almost 8 per cent of the DNA from a group of living humans is derived from other, now disappeared human groups . . . Less than a decade ago palaeoanthropologists thought it unlikely that there was any interbreeding between modern humans and Neanderthals, and perhaps that they never even met each other. Now we know this is *spectacularly* wrong.

Another intriguing piece of data from the paper was derived from mitochondrial DNA extracted from a human tooth from Denisova Cave. Denisova 4, as it is known, is a very well-preserved molar that was discovered in 2000 in the part of the cave we call the South Chamber. Pääbo and his team wondered whether this might be from the same human individual as Denisova 3. Unfortunately, compared with Denisova 3, the preservation of nuclear DNA was not nearly as good. Mitochondrial DNA, however, was eventually extracted from the tooth with coverage of 58×, or every position in the mtDNA genome being covered an average of fifty-eight times. The sequence differed in two positions when compared with the mtDNA sequence of Denisova 3, so it was not from the same person, but the fact that it was so similar suggested that it probably was from the same population of people, with not too great a separation in terms of time. The DNA evidence suggests the two shared a common ancestor less than 7,500 years before they both lived. This demonstrated that with the Denisovans, then, we really were dealing with a human population.

Analysis of the tiny fragment of bone that is Denisova 3 literally rewrote the history of humanity in a matter of months during 2010. The small size of the bone and the reduced presence of fragments of Y chromosome DNA meant that the genetics team could determine that Denisova 3 was a bone from a young teenage girl. Her beautiful gift to us was to show that alongside us and Neanderthals in mainland Eurasia there had lived another group of previously invisible people. In addition, her genome told us that some of the modern human family had inherited some Denisovan DNA and that those people live today in Melanesia and Australia. It becomes apparent from this, then, that when we talk about 'modern human origins', we clearly need to widen the term to include not just our own specific evolution, but the admixture that has come from other groups, previously considered to be more 'other', than 'us' – or, in the case of Denisovans before 2010, unknown.

The Denisovan genetic data were so comprehensive in comparison with the scarce skeletal remains we have of the group – just four tiny remains – that one journalist referred to Denisovans as 'the genome in search of a fossil'. In the next chapter, I turn to the efforts aimed at finding more Denisovan remains in an attempt to put more flesh on these fragmentary bones.

7. Where are the Fossil Remains?

How can we find out what Denisovans looked like if we only have tiny fragments of them? The Denisova 3 genome provides information about the phenotype of the girl from Denisova, but to find out more about the basic skeletal and cranial morphology of these people we really need new fossil remains.

There are two possible places to look. First, we can excavate new archaeological sites. This takes time and effort, but is definitely a profitable route and one that many archaeologists are currently engaged in. Second, we can examine museum and university collections for human remains that have yet to be identified and linked with a particular human species. Is it possible that amongst the skeletal remains housed in different collections, known hitherto only as

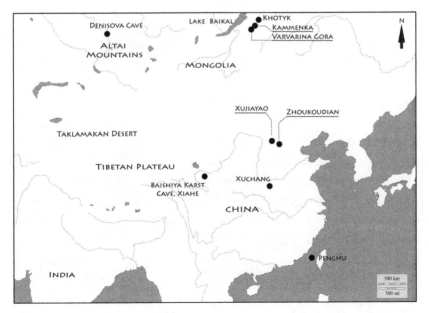

Figure 15 Far Eastern sites and locations.

Homo sp., or 'Archaic *Homo*', there could be Denisovans, or even other types of human lurking?

A starting point is to examine the meagre remains of Denisovans that we have at present, so that we might compare their morphology or shape against extant fossils in eastern Asia. By 2019 the Denisovan specimens we had were the pinky bone (Denisova 3), the second deciduous molar tooth excavated from the base of the early excavations in the cave in 1984 (Denisova 2), the very large upper left molar found in 2000 (Denisova 4) and another upper left molar fragment (Denisova 8, found in 2010).[1]

The most informative specimen is the Denisova 4 tooth, because it is virtually intact and teeth are often amongst the first items used in the diagnosis of a human species; they often reveal distinct species differences when subjected to shape and morphological statistical analysis. It is tempting to think that Denisovan physical remains would be similar to those of Neanderthals, based on their more recent split from a common ancestor around 400,000 years ago,[2] but this is not necessarily so. Take the Denisova 3 pinky bone, for example. After the two fragments of the specimen were reunited it was possible to compare its physical proportions with other hominin phalanges. When forty-five Neanderthals and thirty-one ancient (30–110,000 years ago) or recent modern humans were compared against Denisova 3, it plotted closest to the anatomically modern humans. It was particularly slender and gracile, and surprisingly modern human-like in its appearance[3] – quite different from Neanderthals. This might suggest that the thicker finger bones of Neanderthals evolved in a different direction following their split from Denisovans.* The similarities between modern humans and this Denisovan in the finger proportions suggest we should therefore be cautious when we try to identify possible new Denisovan skeletal remains through comparison with small elements such as this.

---

* Neanderthal fingers appear quite specialized and part of their more generally sturdier phenotype, in this case incorporating a very strong grip with wide and robust fingertips.

Denisova 4 is *most likely* to be a third upper molar, or wisdom tooth (termed the M^3),* but because it is a single specimen, not found within its original jaw and dental arcade, it is difficult to be sure. Generally, from the time of *Homo erectus*, after 1.6 million years ago, human molars have become steadily smaller. Our molar sizes decrease from first to third molar, but in other hominins, the reverse is the case. Australopithecines and early *Homo* tooth sizes *increase* from the M^1 to the M^3. Interestingly, despite it being a third molar, its large size suggested to Bence Viola (who led the study of this tooth) that it must have been the largest molar in the tooth row. He therefore compared Denisova 4 with M^3 teeth from other hominins and, for good measure against M^2 teeth as well, just in case it wasn't an M^3.[4] He measured two angles on the tooth, one lengthwise and one across its widest cross-section. The M^3 was extremely different from any Neanderthal or modern human tooth. It plotted closest to Australopithecines and *Homo habilis*, significantly older African hominins dating to more than 1.5 million years ago. The M^2 comparisons were similar: Denisova 4 was much larger than Neanderthals or modern humans and tended to fit with much older hominins and at the top of the range exhibited by *Homo erectus*.[5] The shape of the roots and crown too were very different. The Denisovan tooth was miles away from its expected closest relatives in the hominin record. How can this be?

Neanderthal teeth begin to show distinct characteristics very early in the evolution of their lineage. One can see this in fossils from the key site of Sima de los Huesos, or 'Pit of the Bones' (near Burgos in northern Spain), which dates to more than 400,000 years ago.[6] It is curious that Denisova 4 appears so very different from Neanderthals when they both share a common ancestor whose estimated age is not significantly different from the age of the Sima fossils.

If the tooth is so distinct, then are there any archaeological specimens in East Asia that are similar to it?

One possibility is some human material that comes from a site

---

* Teeth are termed M for molar, I for incisor, P for premolar and C for canine, and their position is given with a subscript or superscript to denote whether they are upper or lower. So an M^1 is an upper first molar while a M_3 is a lower third molar. Lower case letters are used for deciduous teeth.

called Xujiayao, in northern China. In the late 1970s, excavations there revealed an upper jaw with six teeth still in place, as well as some other isolated teeth and sixteen skull fragments.[7] The age of the site has been difficult to determine, but the latest work suggests the human remains to be more than 140,000 years old.[8] Analysis of the teeth shows that the molars are also very large and robust, with a primitive shape, rather like Denisova 4. The front teeth, the incisor and canine, however, are more like those of Neanderthals. One specialist has suggested that the overall morphology of the teeth 'preserves the heritage of a "mysterious" and primitive hominin',[9] the most likely candidates for which are the more ancient hominins of the Middle Pleistocene of East Asia. This is a term describing hominins that are not yet us, because they retain aspects of archaic morphology (including a skull shape that is less globular than ours and teeth larger than ours), but at the same time are not Neanderthals. Some features of the Xujiayao humans evoke Neanderthals. For example, in the very important inner ear bones, the bony labyrinth is very similar to that of Neanderthals. Other features align more with earlier Pleistocene humans of East Asia.

Taken together, perhaps the word that best describes the Xujiayao hominin material is 'mosaic': a mosaic of features never before seen in Eurasia. While acknowledging that they may well be the remains of a Denisovan, the unusual features and mixed traits have led some to suggest that the Xujiayao remains could perhaps come from a form of hybrid human.[10] We will consider this in more detail later but, as we shall see, it does make sense when we look at what we see in genetic hybrids amongst other animals and other primates. They often display evidence for excessive size: larger teeth, larger skull shapes and so on – which is interesting when we think back to the extreme size of Denisova 4.

There are some parallels between the Denisova 4 tooth and Xujiayao, then, which could turn out to be important, but are not enough to provide the smoking gun in terms of linking Xujiayao to the Denisovan lineage.

There are human remains from other sites that might be informative. Lingjing is near Xuchang in the northern Chinese province of Henan. In 2007 archaeologists there excavated a yellow piece of rounded skullcap that was found eroding out from sediments that also contained a group of beautiful quartz tools.[11] By 2014, after seven further periods of excavation, the team had excavated two crania, lacking jaws and the frontal bones of the face. The age estimated for the layer within which the Xuchang human remains were found is 105–125,000 years.[12] The specimens again show a mix of features: some of them evoke similarities with other human remains in East Asia, while others are remarkably similar to Neanderthals. The oval outline of the skull when viewed from behind, for example, is very evocative of Neanderthal skull shape. The estimated brain size of 1,800cc is also extremely large,[13] in fact it is bigger than any Neanderthal and of a size practically never seen in a modern human.

Could DNA help us to diagnose whether the Xuchang remains are Denisovan?

Qiaomei Fu, of the Institute of Vertebrate Palaeontology and Palaeoanthropology (IVPP) in Beijing (whom we met in Chapter 6 finding Denisova 3 with Johannes Krause), took on the sampling and genetic analysis of the Xuchang crania. Sadly, however, she and her team failed to extract enough useful DNA. Currently, then, we simply cannot be sure of their taxonomic status. The curious mosaic implies that they may come from a population that has undergone some admixture, but, for now, we just do not know.

The fragmentary nature of the remains we have from Denisova Cave, along with a dearth of comparable specimens and their disappointing lack of DNA, has made identifying further Denisovan remains elusive. But this isn't a dead end. One possible way to shed light on the Denisovans and what they may have looked like has come from the field of computational biology and the emerging science of epigenetics.

Epigenetics refers to the observation that phenotypic expression, or the characteristics and morphology of an organism, can change *without* actual alterations to the DNA sequence of the organism. It sounds counter-intuitive, but it revolves around chemical changes that can switch off or modify certain genes.

Genes can be affected by the addition of a chemical 'methyl group' to bases in the DNA sequence. The methyl group comprises one

carbon and three hydrogen atoms, and when methylation happens
it can change the function of a gene. Genes can be switched off and
on by methylation; their expression can be altered by these chemical
knobs. If one can work out the methylation patterns, it might provide clues to phenotypic expression.

There is just one problem. After the death of the organism, and the
degradation of its DNA and other biomolecules in archaeological
sites, the methyl groups also degrade and disappear. With them goes
the potential opportunity to reconstruct appearance. There are some
notable exceptions, however, for example a 26,000-year-old permafrost bison was so well preserved that its ancient methylation patterns
could be reconstructed.[14] But usually it is just not possible, and so far,
no Denisovan or other ancient human relative has ever been found in
permafrost.

So how, then, can we extract this information?

An Israeli team has worked on an answer, which is to analyse *proxy*
signals for methylation.[15] The key is to look at differences in the
chemistry of the nucleotide base cytosine over time. In Chapter 5, we
briefly mentioned cytosine decaying to uracil over time (a process
called 'deamination'[16]) and how this is used to isolate truly ancient
DNA fragments. In this instance, the cytosine is unmethylated, or
has no added methyl group. Cytosine that is *methylated*, on the other
hand, decays to thymine, rather than uracil. Positions on the genome
which have cytosines that were methylated prior to death and decay
therefore ought to possess a higher fraction of thymine compared
with unmethylated positions. From this it can be deduced that the
C→T ratio acts as a proxy for methylation in ancient DNA samples.

To test the predictive power of their method, the Israeli team initially examined the methylation patterns on the chimpanzee genome.
They were able to test the strength of their approach by comparing
their predicted phenotypes, based on the so-called methylation maps
produced, with the known physical appearance of chimps. They also
explored the method by applying it to Neanderthals, comparing their
physical appearance against Human Phenotype Ontology, a database
of genes that influence changes in anatomy when they are turned off
or missing. Initial tests showed that they could reconstruct certain

physical traits in both groups to more than 85 per cent accuracy.[17] This seemed very encouraging.

Then they analysed the equivalent Denisovan methylation map, focusing on those parts that covered sets of genes underlying particular traits. They paid attention to areas that indicated a directionality in the expression of a trait, for example away from a Neanderthal or towards a modern human. In this way they built a comparative picture of the possible appearance of a Denisovan.

So, what would a Denisovan look like?

Twenty-one features identified by the research team appeared to be shared with Neanderthals. They predict that a low cranium is expected, for example. A wide pelvis, a robust jaw, a flatter forehead and a large ribcage are also likely to be characteristics of Denisovans.

Some features were distinct to Denisovans. These included greater facial protrusion, larger shoulder blades, an elongated dental arch, an enlarged mandibular condyle (the protuberance at the top of the jawbone where it attaches to the skull just forward of the ear), and a greater width in the face. They also appear to have different bone mineral density and different timings in parts of their skeletal maturation.[18]

Interestingly, Denisovans appear closer to the morphology of modern humans in some traits. These include the premature loss of permanent teeth and the width of the front of the jaw (interesting, bearing in mind what we saw of the great size of the Denisova 4 molar at the rear of the jaw).

Our attempts to compare actual Denisovan remains with other remains of hominins from archaeological sites in East Asia failed to elucidate a clear picture, but they may make sense in the light of these epigenetic predictions. While the majority of the features of Denisovans echo Neanderthals, others are more like modern humans while still others are like neither.

Of course, this is the first step in the application of this methodology, and there are some acknowledged weaknesses. One is that only one Denisovan genome is being analysed. A population is being modelled on the basis of one individual genome. One critic said that if you were to find a single *Homo sapiens* fossil and it turned out to be the remains of an NBA basketball player, you might well conclude that *Homo sapiens* were all seven feet tall. Humorous, but not really a fair criticism, because, as we have seen, the nuclear genome contains information that derives from more than one person; it is the reflection of a wider and deeper population history. The analysis is also lacking in quantification, because it only tells us about *direction* of difference rather than providing hard numerical confidence. However, in the absence of more solid skeletal evidence for Denisovans, it is a significant and rather exciting step forward.

The authors were able to create an artist's interpretation of their reconstruction to show us what a Denisovan might have looked like. Often, artists' reconstructions can be more speculative than accurate, but in this case the artist, Maayan Harel, was working from a 3D model reconstruction based on the data from the epigenetic analysis, and made the image of a Denisovan woman from that. I get a sense from this picture of her closeness to us and the similarities that we share with these people, rather than the differences. For the first time we can gaze on this long-disappeared member of the human family and look at a person, rather than simply a small assortment of fragmented bones, and for this I am thankful.

When we turn the developing power of this epigenetic tool on some of the specimens we have considered as possible Denisovans from China, such as Xuchang, what does it tell us? Intriguingly, the analysis suggests that many of the traits identified for Denisovans are present in Xuchang, with one exception. Epigenetic predictions therefore would suggest these crania are more likely to be Denisovans.

This new approach will surely play a larger role in future as more fossil remains of Denisovans are recovered and our understanding of the functional basis and phenotypic expression of different hominin genes improves.[19] The key term here is 'more fossil remains', and by this I mean remains that are large enough to be able to be diagnosed by epigenetic approaches. Never before have we had such an urgent need for new human remains. For much of the last ten years many of us working in the field of late human evolution have anxiously waited for any fossil discovery that could herald a new Denisovan, particularly from a new site or region and remains that are morphologically informative.

Then suddenly, on 1 May 2019, a new discovery was published in *Nature*. For the first time, it seemed, a Denisovan had been found from a site other than Denisova Cave.

One day in 1980, in a holy cave 3,280m above sea-level on the Tibetan plateau in Gansu, eastern China, a monk was praying. He took a handful of cave soil as he turned to leave, as is the Buddhist custom there. It was then that he noticed something unusual. It looked like a partial jaw, human-like. Baishiya Karst Cave in Xiahe had yielded many animal bones, so it was not entirely surprising, but there was something about this bone that caught his eye. The monk (whose name is not known to us) decided to take the bone to the late Sixth Gung-Thang Living Buddha, the leader of his monastery. As a highly educated man, he knew it was potentially special, so he kept it carefully in his house.

More than a decade passed. In the 1990s an environmental scientist from Lanzhou University called Guangrong Dong began to work in the area. Mutual friends introduced him to the Living Buddha and the subject of the interesting mandible found at Baishiya came up. Guangrong Dong agreed to take the bone back to his university to study it, but overwork and a lack of expertise in human physical anthropology meant he didn't manage to turn serious attention to it.

It was not until 2016 that two other researchers, Fahu Chen of the Institute of Tibetan Plateau Research and archaeologist Dongju Zhang of Lanzhou University heard about the mysterious jawbone and offered to work on figuring out what it was. They interviewed the monks at Baishiya Karst Cave to try to find out where the jawbone had come from in the cave. None of them had any idea.[20] It wasn't until palaeoanthropologist Jean-Jacques Hublin of the Max Planck Institute in Leipzig was contacted to see whether he could provide some expert assistance that light began to be shone on the mysterious fossil bone. As soon as Jean-Jacques saw photographs of the Xiahe specimen he immediately knew its possible significance, and arranged a visit to see the jawbone first hand.

Parts of it were covered in encrusting carbonates that made it difficult to see the entire surface. Using micro-CT scanning technology, Jean-Jacques's team were able to remove this virtually to reveal its shape. Because the bone was broken in half, they built a complete virtual 3D reconstruction by 'mirroring' the jaw (later, using 3D scanning software, I was able to download and 3D print the specimen to look at its shape for myself).[21] The team removed tiny amounts of the carbonate from various areas of the bone for dating using uranium–thorium methods (see the footnote on p. 20 for a description of how this method works). Three consistent age estimates were obtained, indicating the sample was from at least 160,000 years ago. This is the earliest evidence for a human presence on the Tibetan plateau.

The Xiahe jawbone itself is moderately large, but the teeth are *extremely* large when compared with other hominins. Although there are only two remaining teeth in the mandible, the M_1 tooth (the lower first molar), is at the higher end of virtually all other Middle Pleistocene samples in the measurements that were taken, while the M_2 sits at the higher end of the distribution recorded by *Homo erectus*. The similarities with the Denisova 4 tooth are obvious.

But what kind of human was Xiahe?

Sadly, once again, attempts to extract original DNA from the jaw failed, so, in the absence of DNA, the researchers turned to a very exciting and emerging field called proteomics.[22]

Proteomics is the study of proteins. Our proteins are built based on instructions given by genes. There are around 20,000 proteins in humans that are only now being explored with the advent of new mass spectrometers that can sequence them. A method called LC-MS/MS* enables scientists to identify the sequence of proteins.†

---

* Liquid chromatography-mass spectrometry/mass spectrometry.
† Initially, proteins and amino acids are separated from one another using liquid chromatography and then each is given an electrical charge, which allows scientists to alter their trajectory when passed by a magnetic field at high speed. Larger molecules are deflected to a smaller extent than smaller ones and therefore can be separated and measured, or detected, according to their mass. This enables their abundance to be determined. A second mass spectrometer can be used to further determine the mass of larger particles, which can be broken up into smaller-sized particles and measured.

What is very exciting is that proteins tend to survive for longer in the archaeological record compared with DNA. Considerably longer. Currently, the record for the oldest extracted ancient DNA is held by an ancient horse bone found in permafrost in the Yukon of western Canada, dating to between 560,000 and 780,000 years ago.[23] In contrast, protein in ostrich eggshell has been attested at the site of Laetoli in East Africa at 3.8 million years.[24] Ancient proteins may survive for even longer; the most controversial is the claim by the scientist Mary Schweitzer of protein surviving in 80 million-year-old dinosaur bones.[25]

One problem with proteins is that they are much less informative than DNA. Remember the human nuclear genome as a set of 3,200 books? The equivalent for the proteome would be a glossy magazine, and a rather short one at that. There just isn't the same level of diversity in protein sequences compared with DNA.

In the case of modern humans, Denisovans and Neanderthals, we know that there are demonstrable differences between their genomes. Since proteins are coded for by genes, the question is, are there any differences between the proteins in these groups that could facilitate identification to species level?

Gradually, evidence is emerging that the answer is yes. Small but identifiable differences have been found in single amino acid positions along extracted proteomes, and proteomics is now beginning to be used to place different primates and hominins in evolutionary trees, rather like DNA.[26]

The research group extracted a sample from the bone to see whether proteins were sufficiently well preserved to extract proteomic information. The bone was poorly preserved, but there was more luck with one of the teeth, which yielded degraded and therefore ancient proteins, able to be separated from modern protein contaminants. The presence of two so-called single amino acid polymorphisms (SAPs) that differed between Denisovans, Neanderthals and modern humans, enabled them to place the Xiahe specimen along with other previously analysed samples in a family tree.

The closest neighbour on the tree was Denisova 3 . . .

The team was able to conclude on this basis that the Xiahe mandible from Baishiya Cave was likely to be Denisovan.[27] This makes it the first Denisovan to be found outside Denisova Cave, which is truly exciting. Gradually we are starting to take the first tentative steps towards working out where, geographically, Denisovans may have lived and dispersed into.

As scientists it is important to be cautious, however, and remember that this work was based on only two amino acid positions in the proteome, of which one matched the Denisovan. It is a thin indication compared with genomic evidence. With ancient DNA we usually have much more to support the identification. DNA enables the different species to be distinguished since the mutations that build up through time are mostly neutral. In proteins, changes usually occur because they are under selection. This means that they are much more likely to arise independently in different populations and species. Ancient proteomics is a very new research area and the science is still developing, so we must tread carefully. That said, it seems that this approach has the potential to provide ground-breaking information, particularly in cases like Xiahe where DNA does not survive and in the case of extremely ancient fossils which are too old for DNA preservation. In addition, it is worth pointing out that the shape and anthropological analysis of the specimen seem to support Xiahe being Denisovan regardless of the proteomic work.

Work continues at Baishiya. The cave itself is extraordinarily long; it stretches more than a kilometre into the hillside. At its entrance it is 10m high and 20m wide. There have been surveys and test excavations since 2011 that have resulted in the recovery of a range of stone tools as well as bones with evidence for human modification, including cut marks. My laboratory is currently dating bones from the site to establish the chronology more reliably. Perhaps this fascinating place will provide more evidence for Denisovans and shed light on the intriguing adaptation of these people to living at higher altitudes. Its age suggests that Denisovans had been living in East Asia for a considerable length of time. But how and when did the human family move into the difficult Tibetan terrain? What kind of adaptation and technology did they have to survive? When did they develop the ability to cope with the stresses of high-altitude living? We will explore this in more detail in Chapter 16.

The Xiahe specimen, and its Denisovan diagnosis, once again enable us to go back and explore the existing fossil record to compare its features against other similar bones from the East Asian archaeological record.

One specimen that now appears interesting is Penghu 1.

In February 2008 a man called Kun-Yu Tsai was out shopping in his home town of Tainan City in Taiwan. He was browsing in an antique shop when something caught his eye. It was an ancient-looking bone. When he quizzed the shop owner about it he was told that the bone, a jaw or mandible, had come to him via a local fisherman. The fisherman had been working on a boat in the strait between Taiwan and a cluster of tiny islands offshore called the Penghu Islands. Above an oceanic trench around 25km offshore he had pulled up his net and noticed that, along with the fish, some bones were present as by-catch. The fisherman had taken some of them to the man in the antique shop and sold them. The jawbone was put on prominent display by the shopkeeper, hoping to catch the eye of someone interested in something a little out of the ordinary.

Mr Tsai liked the look of it. He purchased it and took it home. The surface was covered in marine concretions, so he cleaned it up a little. To his untrained eye it looked interesting and he wanted to know more. What exactly was it? How old was it?

In 2009 he took some photographs and sent them to scientists at the National Museum of Natural Science in Taichung City.* They recognized that the mandible looked rather archaic and possibly quite old.† Eventually, they were able to examine and work on it first-hand.

---

* Chun-Hsiang Chang and a Japanese colleague from Kyoto called Masanaru Takai, who had previously worked on fossil monkey remains in Taiwan. Both of them then worked on the Penghu specimen.

† Dating of the Penghu specimen is, not surprisingly, imprecise due to the nature of its discovery. It is thought that it probably dates to between either 10,000 and 70,000 or 130,000 and 190,000 years old, based on estimates of sea-level changes around the coast of Taiwan. The earlier age range is the more likely one.

A glance at Penghu 1 reveals the important absence of a chin. It is similar to the Xiahe specimen, both in its relative size and in aspects of tooth size and shape. Both have a missing third molar. This is a rare occurrence in the fossil record. Both have very large molar crowns. In addition, and much more significantly, they both contain a second molar that has three roots. In non-Asian *Homo sapiens* this is very rare, seen in less than 3.5 per cent of people, but in Asian people and Native Americans it is much more common: up to 40 per cent of people have it.[28] Its presence means it is likely to be a very ancient trait. Prior to this, the earliest evidence for three-rooted molars in humans was found in a modern human jaw in the Philippine site of Tabon, although the dating is not precise.[29]* If Xiahe is a Denisovan, this feature is almost certainly derived in modern humans, and comes from genes introgressed from Denisovans, that is, by interbreeding.[30] This suggests that gene flow from archaic populations to modern humans may explain some of the curious dental features that are found in some modern populations and not others.

It looks as if the Denisovan family of ancient bones may now extend to the Penghu specimen.

The little jaw is now housed in the National Museum of Natural Science, Taiwan,[31] thanks to the financial support and encouragement of a local businessman called Ted Chen. Perhaps, if the latest scientific methods can be applied to it, we may find out more about it in future.

In 2016 a section in the South Chamber at Denisova Cave collapsed, and while the team was cleaning up the sediment, they found two pieces of bone that belonged to the parietal or side sections of a human cranium. The fragments were thick. I saw a cast of them in Novosibirsk the following year and it was impressive; I have never seen or handled a thicker piece of skull bone. It was identified as Denisovan using ancient DNA and given the identifier Denisova 13.[32] It is the biggest piece of Denisovan bone so far found at the eponymous site.

If we take the three teeth from Denisova Cave and the pinky bone,

* Between 9,000 and 47,000 years old.

the Denisova 13 cranial fragment, as well as the Xiahe jawbone, that brings to six the number of identified Denisovan bones in the world. As we expected, it looks as though there are already Denisovan remains in the previous fossil record of East Asia waiting to be redis-covered. Penghu and Xujiayao, as well as Xuchang, fell into this category, although – so far – there is no DNA evidence to corrobor-ate this. It is a tiny group of specimens, but it is slowly getting bigger, and with each new bone we know more. Things move fast in the world of palaeoanthropology; this time next week it could all be different.

If we cannot find other bones in the extant fossil record, then we need to find alternative approaches. In the next chapter I want to explore how exciting new bioarchaeological methods are beginning to play an increasingly important role in finding tiny fragments of bone that can then be genetically analysed and dated. One of these fragments, the discovery of which I related in the opening pages of this book, has caused a sensation in the world of palaeoanthropology. Discovering it was one of the highlights of my scientific career.

8.   Finding Needles in Haystacks

The conundrum of how to find more Denisovan remains, and from where, was a nagging problem in the back of my mind. In July 2014 I went to Denisova Cave with my research collaborator (and wife) Katerina Douka for a meeting. As usual I experienced the joy and exhilaration of being in the Altai Mountains again and feeling at one with nature and the beautiful surroundings at the site: the soothing babble of the Anui river, the clear fresh air and, of course, the unparalleled archaeology and camaraderie of the team. These mini-conference events, held every three or four years, had an interesting line-up of the great and the good from the field of palaeoanthropology. Anatoly Derevianko, Michael Shunkov, Svante Pääbo, Bence Viola and others were there, along with some interesting students and researchers working in diverse areas of archaeology and genetics whom I hadn't met before.

Katerina and I settled into one of the nice chalets in the Denisova camp. Over the first couple of days at the site, across various conversations, chats over drinks and in late-evening discussions together, a beautiful idea began to dawn in our minds.

It occurred to us that the massive challenge and problem of Denisova Cave was that 95 per cent of the bones found in the site are fragmented, probably due to animal scavenging. This means that the vast majority cannot be identified as either animal or human, still less to a species or genus. They just sit collected in large plastic bags in dusty storerooms, having been excavated, cleaned and archived. They are, to all intents and purposes, worthless to archaeology. Yet amongst this detritus must surely lurk human bones, perhaps – like Denisova 3, the Denisovan pinky, or Denisova 5, the so-called Altai Neanderthal – just a few centimetres long but packed with DNA and crucial information about the past occupants of the site. All we had to do was find them. But how?

The answer lay in a novel, recently developed technique called ZooMS: zooarchaeology by mass spectrometry. Different animal species have very slightly different sequences of proteins (or peptides) in their bone collagen. When measured using a mass spectrometer these peptide sequences provide a kind of molecular fingerprint that enables the bone to be identified to species or genus level by comparing its fingerprint to others in a library of bones of known species. The method was originally developed by Matthew Collins and his student Michael Buckley at the University of York.[1] Katerina and I realized that it might be possible to use this method to screen thousands of bones for only those with human peptide sequences, and thereby, with luck, add to the tiny corpus of human bones from the site.

We weren't optimistic. It is rare in the Palaeolithic to find human remains; almost all the bones discovered are from animals. Occasionally we have skeletal remains or the odd tooth, but these are a tiny minority compared with the animal remains that humans hunt and predators brought to caves. Would sifting through all those bone fragments really be worthwhile? Finally, we raised the idea at a small dinner party hosted by Anatoly Derevianko. At some point during the vodka-fuelled evening we pitched our idea. Anatoly and Michael Shunkov were supportive. Svante Pääbo was also very interested; he could see immediately that it might yield more human bones from the site that could be amenable to genetic sequencing, given the high level of biomolecular preservation there. We were given the green light. Anatoly agreed that he and Michael would bring us a bag of bone fragments for us to start working on when we next met, which would be at a conference in Burgos, Spain later in the year.

In the case of hominins, ZooMS can tell us if a bone is from the Hominidae, but nothing more.* In other words, a gorilla, a Neanderthal, a Denisovan, an orangutan and you and I will have the same

---

* The Hominidae refers to the family that includes four genera of great apes: *Pongo* (orangutan), *Gorilla*, *Pan* (the common chimpanzee and the bonobo) and *Homo*.

sequence of peptides on the mass spectrometer. What is needed to distinguish things further, and identify the sample properly, is DNA extraction and sequencing. For this we would need to rely on Svante and his world-leading research group.

This visit to Denisova was the first occasion I'd spent much time with Svante. We had met very briefly back in the early 2000s in a pub in Oxford with a larger group of people, but at Denisova, in the smallish project meetings, you really rub shoulders with your colleagues and get to know them. I'd spent most of a decade travelling to take samples from sites and museums for various dating projects. Every time I sampled a Neanderthal bone for radiocarbon dating, I always asked whether the curators would be interested in sequencing for DNA. The answer would invariably come back 'this sample is in Leipzig with Svante'. I got used to disappointment. It was because they were the specialists in Neanderthals, and later Denisovans, of course. And the best.

Svante began the field of ancient DNA almost single-handedly in

Figure 16 Svante Pääbo (left) and the author at Denisova Cave in 2014.

the 1980s in the laboratory of Allan Wilson in Berkeley. Now in Leipzig, the department he leads at the Max Planck Institute is the world leader in ancient genomics. He's a tall and rather gangly Swede, who says 'very cool' a lot and has a great sense of humour. I got on well with him from the start and we are now very close friends and collaborators. He's razor sharp, and geeky in a good way.*

Svante's department has an annual get-together to discuss research. I was lucky enough to be invited to join them as a guest once at Pula in Croatia. It was a great insight into how his research interests and large team mesh. Svante is the leader and directs a hugely talented team of established researchers and bright young students.

The idea to try ZooMS on the Denisova samples dovetailed nicely with my receiving a large research grant from the European Research Council.† We were able to use some of this money to run several thousand ZooMS samples, so, once back in Oxford, we began the task of preparing the small fragments of bone for analysis. This first stage is the most time consuming. It involves the sawing of a small fragment of bone from each sample, around 20mg in weight, and then labelling a small plastic tube to house the material as well as a plastic bag for the original bone sample.

Following this, some chemistry is required.

First, we extract the protein 'collagen' from the bone by dissolving most of the bone away using a dilute acid. The collagen then needs to be broken up into single peptides. For this we use an ingenious

---

* His students told me a funny story about an award he was given in Los Angeles in 2016, one of the $3 million Breakthrough Prizes, known as the 'Oscars of Science', which are awarded annually to the world's best scientists. Svante came back to the lab with a photograph of him being presented with the award by a very attractive blonde woman. Svante had no idea who she was until his students told him that she was the mega-selling pop star Christina Aguilera.

† The ERC provided a €2.5 million grant, an unusually large amount of money, which is part of the European Union approach to funding blue-sky, high-risk/high-gain research. My project was called 'PalaeoChron', standing for 'Palaeolithic Chronology'. I was able to employ several post-doctoral researchers and students over six years.

method which involves adding an enzyme called trypsin to the collagen solution. It cuts the collagen into individual peptides with surgical precision. (In fact, trypsin does the same job in our gut, where it helps us to digest proteins. The next time you are eating a large steak you will be undertaking the same process of enzymatic digestion that we do in the laboratory to find human bones in archaeological sites.) Once the peptides are separated a chemical is added that allows them later to harden and crystallize; then tiny amounts of this material are spotted onto steel plates using a multichannel pipette, one after the other in their hundreds. This plate is then put into a mass spectrometer.

A laser is used to excite the peptides and give them a charge. The particles thus charged then travel along a short tube towards a detector; the smaller the fragment of peptide, the faster the journey; the larger ones take longer. In a very short time, the masses of all of the peptide particles in the sample can be measured. Then, each can be compared against a library of previously measured known bones, and thereby identified either to species or, more often, to genus.

The first preparation stage of the work is laborious and requires a dedicated worker. Katerina and I wondered whether one of our students might be interested in taking on the project for one of their dissertation topics. Fortunately for us, that year we had a strong cohort of Masters students and one, Samantha Brown, a bubbly and motivated Australian, volunteered to work on the project for her MSc.

Sam started sawing bones. She disappeared for days in the drilling room of our lab in Oxford and by the end of January she had taken 10mg samples from just over 700 bones for analysis. In mid-February 2015 she made the train trip up to the Manchester laboratory run by Mike Buckley, where she worked to analyse the samples and learnt how to identify the peptide sequences.

A week later, an analysis of the spectra revealed that the bones we had sampled belonged to bear, cow, deer, dog, fox, goat, horse, hyena, mammoth, mouse, rabbit, reindeer, woolly rhino and sheep. But no Hominidae. Our attempt had failed. Lots of hard work, but

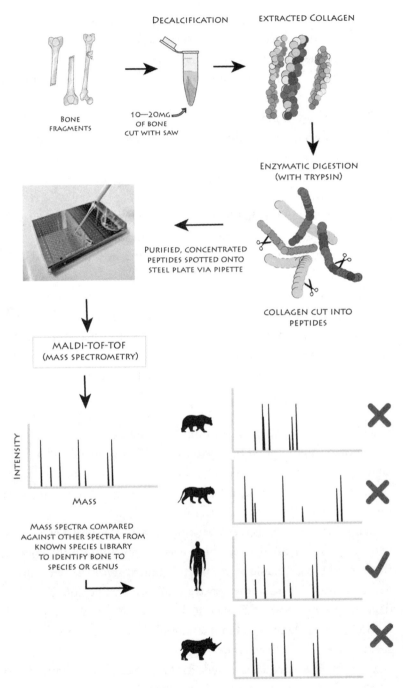

Figure 17 The method of using ZooMS to identify human bones.

no humans. I resigned myself to the fact that we were truly looking for a needle in a haystack – well, bonestack. It hadn't worked, but at least we'd tried. I felt we'd let Sam down. She had put in a lot of work and obtained some data but not the 'money-shot' result we'd hoped for.

We arranged to meet Sam to discuss the results. I said that she still had a lot of good data in the form of the animal bone identifications, and that this was probably sufficient to form the basis of her dissertation. She could relax a bit, take her time and write up her thesis well in time for the deadline.

It was at this point that Sam proved her great worth. She told me no, she didn't want to do that, she wanted to continue, to saw up more samples and press on in the hope that a hominin could still be found. Could she?

There was room in the budget, so why not? The only problem was that we'd run out of bones, so we sent Sam to Russia and a few days later she returned with another large bag of Denisova bone fragments courtesy of our Russian colleagues. Back to the drilling room she went. Three weeks later she had a total of 1,308 samples sawn up and labelled ready to go.

By this time Sam was acquiring something of a reputation for dogged persistence. People were joking that she was practically living in the laboratory, sampling bone after bone after bone. She claimed she enjoyed it, but it is very repetitive work. I hoped so much for success. *Just one bone would be enough*, I didn't care how big; anything to demonstrate this method would actually work.

Sampling done, Sam went back up to Manchester to run another 780 analyses. Unfortunately, there was a problem with the computer there, so she wasn't able to return to Oxford with the data; instead, Mike promised to send the spectra on as soon as he could. The next day, on the evening of Friday 19 June, he was reviewing the data before sending them, and having a quick scan of some of the various spectra when he noticed something.

One of the spectra seemed to have the five unique peptide markers that characterize the Hominidae. It must be a hominin! He triple-checked it and then messaged me at 8.09 p.m. with an email headed 'Success – one out of this week's 780!', to tell me the amazing news: we had a hominin bone!

My reply, sent one minute later at 8.10 p.m., conveys a little of the crazy excitement of that moment:

FUCK FUCK FUCK!!!!!
WOW FUCK!!
SHIT!!
REALLY?????????
(i'm trying to keep calm here)> . . .
FUCK.

Apologies for the flowery language, but there are times in science when you basically lose it with sheer excitement and joy. We'd actually found a hominin bone!

I immediately called Sam on her mobile to share the news.

No answer.

More calls.

Dead mobile.

It turned out that Sam was out at a college 'bop' on Friday night and her phone was out of battery. Damn! It would have to wait. I went downstairs to tell Katerina. We danced around the kitchen with joy and opened a bottle of wine.

Next morning Sam finally read the flurry of excited texts I'd sent; not even a crushing hangover could dent her excitement. We'd actually done it; we'd found a hominin amongst thousands!

And so it was that the following Monday, Sam and I met at 9 a.m. to go through the bags of bones she'd sampled so diligently to find the one that corresponded to the hominin spectrum identified. It was coded DC1227. I videoed the moment on my camera as Sam sifted through the hundreds of individually labelled bags. After a couple of minutes, she had the bone.

It was tiny.

Our excitement tempered any feelings of slight disappointment.

Having worked so hard to identify this bone, and both being positive people, we just focused on what we could do with it. We found that the sample came from Layer 12 in the East Chamber at Denisova. That meant it was almost certainly more than 60,000 years old. We needed to treat it carefully – after all, we wanted to extract DNA from it – so handling was minimized. We did, however, want to get an approximate weight, which turned out to be 1.68g. It was denser than it looked. This might augur well for the preservation of bio-molecules like collagen and DNA.

I found it difficult to sleep that night, such was my level of excite-ment; my head was constantly shaking with disbelief.

We decided to obtain a CT scan, which necessitated a trip out to a lab in rural Oxfordshire. By this time Sam had taken to carrying the bone around in a yellow plastic case I had given her, complete with a handle and special sealed clips. I emailed Sam to arrange a time to meet, adding: 'Oh and Sam, don't forget to bring the bone (in its spe-cial case)!'

Sam replied: 'I could never forget Denny (yep, I named it)!'

'I felt like DC1227 was a little cold,' she said later. 'It felt like a new friend that we found . . .'

So true. From then on, our bone was always Denny.

It turned out that the CT scan would be crucial in determining more about this tiny bone and from whom it derived.

The next job was to get the bone genetically sequenced. Sam took her box (with Denny in it) to the Max Planck Institute in Leipzig, where 30.9mg of bone was carefully drilled prior to DNA extraction and sequencing, initially for mtDNA.

We also took some of the bone for radiocarbon dating and carbon and nitrogen isotope analysis. We expected that, being from Layer 12, the sample should date to before 50,000 years, the maximum age reachable with radiocarbon. But we wanted to be sure that the bone was not younger, so we sacrificed some for collagen extraction and dating. Sure enough, a few weeks later it was confirmed: the bone was greater than 49,900 years old, but how much older we did not know.

Around the same time as we were waiting for results, we saw a new and related paper that had been published in the *Journal of Archaeological Science*.[2] Thankfully for us they hadn't identified any hominin bones. We wanted to be the first to prove the method worked and publish a paper, and it would have been disappointing to have been pipped at the post. Clearly, however, others had been thinking along the same lines as us.

On 9 September we received news that Denny had a Neanderthal-like mitochondrial DNA sequence! The Max Planck team had been able to recover an almost complete mitochondrial genome. We had found a Neanderthal, or more correctly, as Svante cautioned his group to say, a hominin with Neanderthal-like mtDNA. Despite being extremely happy, I remember Sam feeling a tiny bit deflated; once we had a hominin bone, we really wanted it to be a Denisovan. A Neanderthal was amazing, but a Denisovan would have been a lot cooler. People we told about the result, colleagues and friends alike, were genuinely amazed that a technique like ZooMS could identify a tiny human bone amongst thousands, from a person who lived tens of thousands of years ago.

The next job was to publish the paper. Denny was given the identifier 'Denisova 11'. We submitted to *Scientific Reports* in December 2015, and the paper was accepted after review and published on 29 March 2016.[3] I decided to accompany the paper with a press release about the bone and how it was found. I thought it was a fantastic story and that lots of people would be interested. I felt the work was a real breakthrough, showing the power of using bioarchaeological methods to search the archaeological record and find tiny fossil remains.

The story died. It was only covered by a tiny number of newspapers, none of the major ones, and no one in the main scientific press. No journalists called to interview us or discuss anything about the paper. Never mind . . .

Denny then underwent nuclear DNA extraction (undertaken by Viviane Slon as part of her doctoral research). This is a very big job so months went by with no news. In May 2017 Katerina and I were in Germany and we popped into the Max Planck to see how work was progressing and grab a coffee. Members of the team there said that they had made good progress and the initial results were exciting but strange, so needed confirmation. When I asked for specifics they swore us to secrecy before telling us that the results seemed to show an unexpected degree of mixing in the DNA, that some of the genome seemed to align with the Denisovan genome and some to the Neanderthal genome. It was, they said, roughly split, around 50–50. They said that the initial evidence suggested that we had a so-called 'genetic F1 hybrid' on our hands. There was no other way of putting it: Denny had two parents, a Denisovan father and a Neanderthal mother.

I wish I had a video of the moment I was told this news; I think my jaw must have dropped open as I sat stunned and wide-eyed. We all were. I was thinking, how could this be? There must be a mistake, a mix-up, some kind of contamination or other error. Later, the extracts were done again and again to make sure. Same result each time. It was an incredible thing to comprehend. We told Sam the 'secret' news, that she may have found a hybrid – not a Neanderthal, but something part-Denisovan after all!

Ultimately, Denny's genome was sequenced to an average of 2.6× coverage. This revealed that she was a female. To figure out her ancestry, her genome was compared against the high-coverage Neanderthal (Denisova 5) and Denisovan (Denisova 3) genomes, along with an African genome (Mbuti) that ought to have no DNA from either group as a control. This analysis showed that at 38.6 per cent of sites on the nuclear genome, Denny matched the Neanderthal genome, and at 42.3 per cent of the sites her alleles matched the Denisovan genome. So, there was indeed an equal split in terms of the contributing DNA. This demonstrated with confidence that she was indeed an F1, a first-generation offspring of two different human groups.

Based on the CT scans, by comparing the cortical thickness with values from other hominins, we could determine that the bone must have come from a girl not younger than thirteen years of age.

Gradually we were putting flesh on this tiny, 2cm length of bone.

The paper describing the results was published online in *Nature* on 22 August 2018, and a couple of weeks later in the paper version of the journal, as the cover article.[4] We had a beautiful cover commissioned by a German artist, Annette Günzel. It comprised an image of two clasping hands, one in blue and one in red, designed to signify the two hominin parents of Denny. In the background was a map of Eurasia. A strand of the DNA double helix threaded along the hands and over the forearms rather like a tattoo.

This time the press release was picked up all over the world. It was literally everywhere. Denny was one of the biggest science stories of the year. Viviane appeared in *Nature* as one of their Top 10 People for 2018. Geneticist Pontus Skoglund said that Denny was 'probably the most fascinating person who's ever had their genome sequenced'.[5]

What a delight to discover all of this and then to share it! Magic! I get tears in my eyes thinking about it all again, I really do.

There are three things about Denny that I want to touch on before we move on. The first concerns understanding the likelihood of finding her and whether, as a first-generation person of two different human groups, she represents an incredible piece of luck, a one-off. People like Denny could not be a common occurrence; if she were then the genomes of Neanderthals and Denisovans would be the same, not different. There had been interbreeding identified between Neanderthals and Denisovans before in their population history. We had known since 2014 that there had been gene flow from the Altai Neanderthal population into Denisovans.[6] So, the two populations were separate, based on the genetics, but I think must have interbred on the occasions that they met. Of course, in the future we might well find more examples of interbreeding and more first- or second-generation offspring, who knows, but I think it is less likely. I think we just got very lucky.[7]

Another intriguing observation from Denny's genome, however, provided more hints about interbreeding and its frequency. At five locations on her nuclear genome there were tracts of predominantly

Neanderthal-only DNA, which showed that Denny's Denisovan father himself had a Neanderthal ancestor (or ancestors) hundreds of generations before. The DNA of the introgressing Neanderthal also disclosed more interesting evidence about the ancestry of Denny's mother. Her DNA indicated that she was from a *different* population of Neanderthals from those who had interbred with Denny's father's ancestors. It turns out that she was more closely related to a 50,000-year-old Neanderthal from Croatia, from the site of Vindija Cave, than she was to other, later Neanderthals from Denisova Cave. This suggested that either western Neanderthals had migrated to the Altai or that Neanderthals from the east had moved to Europe and replaced the Neanderthals there. The genetics therefore provides insights into quite significant population movements in the Neanderthal lineage during this time.

The final thing I want to talk about briefly concerns human remains.

I have taken samples for radiocarbon and stable isotope analysis from hundreds and hundreds of human bones during the course of my career. For decades I prided myself on not being emotionally attached to them, of being someone detached and remote from the ancient individual I was taking samples from. It didn't matter, the age: 100 years, 1,000 years, 10,000 years – for me it was always about the science and not the individual. I have worked on the skeletal remains of Richard III, John Merrick (the so-called 'Elephant Man'), Egyptian pharaohs, bog bodies and mummies, saints and murder victims; I've always maintained a sober and detached face, respectful of the fact that I was working with the remains of real people. But Denny changed this in me. Perhaps it was the intense period of work that it took to discover her, and the rollercoaster ride, but finding out who she was became something bigger and more personal than just science.

As I write this in my office at work, the last tiny remnant of Denny is sitting just behind me wrapped in a plastic bag in its little yellow box, so perhaps I feel her presence more than I otherwise would. Denny was formerly anonymous, as so many human remains are in archaeology, and this tiny bone is perhaps the last surviving fragment of her. But gradually we have been able to bring her back into the light, to breathe life back into her and to reconstruct aspects of her existence. Through this I think that we have begun somehow to honour her memory and give her story to the world. In the process Denny the 'sample', the 'bone', has become Denny the person.

9. The Science of 'When'

'Everything that comes from ancient heathen times floats before us as in a thick fog, in an immeasurable period. We know that it is older than Christianity, but whether it is a matter of years or centuries – or even over a thousand years – older, is the subject of pure guesswork and at best of only probable hypotheses.'

So said Danish antiquarian and philosopher Rasmus Nyerup in 1806. As someone who works in a scientific laboratory that can literally date when things happened in the past, I have often thought how wonderful it would be for Nyerup to know the difference between his plaintive plea to understand the temporal component of human history then and now. How amazing it would be to transport him via time machine from 1806 to today and show him around our laboratory and introduce him to what it does.

In order to further place Denny, and other human remains from Denisova Cave and elsewhere, into a proper perspective, we really need to know *when* they lived. Chronology is crucial. Without a means of ordering past events, determining what happened when becomes impossible. Nothing but Nyerup's thick fog results. We have already heard about the dates from other sites where Denisovans or possible Denisovans have been found: Xuchang, Xiahe, Xujiayao; but what about the age of the finds from Denisova Cave itself?

Before March 1949 archaeologists had to rely on what we term 'relative dating'. The approach was initially centred upon the eastern Mediterranean and, in particular, the historical record of Egypt. By comparing pots and other objects of a known age there, it was possible to infer the ages of similar objects found in other areas of the eastern Mediterranean. Relative dating was restricted, both geographically and temporally. Gradually, this comparative framework was extended more widely across Europe, but it was always relative, not absolute.

The 4th of March 1949, however, marked the publication of a

seismic paper in the journal *Science* and the beginning of a significant change in how we go about dating the past.[1] Written by Willard Libby, a scientist at the University of Chicago who had formerly worked on the Manhattan Project that produced the world's first nuclear weapons, it described a new method that would enable archaeologists around the world to independently date archaeological samples. The only prerequisite was that the material being dated had once to have belonged to a living organism. Charcoal, bone, shell, wood: all could be dated using the new method. This was a revolution in archaeology and a range of other sciences, for which Libby would be awarded the Nobel Prize in Chemistry in 1960.

Radiocarbon dating is almost seventy years old, and it has gone through several revolutions since the development of the technique. One of these is to use particle accelerators to measure the ^{14}C isotope. Radiocarbon is the radioactive isotope of carbon, but it's not present in significant enough amounts to make it dangerous to humans, so we refer to it as a low-level radioactive isotope. To give you an idea of how low-level this is, compared with the majority of carbon in the world, which is in the form of carbon-12, or ^{12}C, for every 1,000,000,000,000 ^{12}C atoms, we have one ^{14}C atom. So our machine has to be sensitive enough to detect at least one atom in a trillion. The accelerator mass spectrometry (AMS) method enables the dating of milligram-sized samples, as small as a quarter of a grain of rice, and quickly. We usually analyse samples for less than twenty minutes each, but in fact the instrument is so advanced that even after forty seconds we get a pretty good idea of the actual age of the sample. We can date material back to the limit of the method, around 50,000 years ago. This is because the half-life of radiocarbon is 5,568 ± 30 years. So, every 5,568 years the amount of ^{14}C in any given material reduces by half. So two half-lives are 11,136 years, three half-lives are 16,704 years, and so on, until by nine half-lives you are back to just over 50,000 years with virtually no ^{14}C left to measure.* Ten half-lives are accepted as being the maximum any radio-isotopic method can reach.

---

* This means that when we date samples of Neanderthals or Denisovans we are measuring at even lower levels of isotopic detection: 1 in 1,000,000,000,000,000,000.

The laboratory I run at the University of Oxford is one of the first accelerator laboratories in the world, and the only facility that specializes in dating archaeological samples. We almost never date samples that are not archaeological. This is because, unlike almost all other labs, we are based in an archaeology department, interestingly founded just after Libby's discovery, and inspired by it. Radiocarbon essentially gave birth to the field of archaeological science.

When I first came to Oxford in 2001 one of the first people I met was a Palaeolithic archaeologist called Roger Jacobi, whom we encountered earlier as the scientist who would lick bones to see if they had been glued. Roger had been working on the British Palaeolithic since he was a young student, and had personally submitted dozens of samples from Neanderthal and early modern human sites to the laboratory for radiocarbon dating – more, in fact, than any other person. He was beginning to realize, however, that dating samples between 30,000 and 50,000 years of age was trickier than people had first thought. He was right. The main challenge is that with increasing age come decreasing levels of ^{14}C to measure. At 30,000 years ago, there is only 3 per cent of the ^{14}C that we have today. By 40,000 it's 0.7 per cent and by 50,000 it is only 0.1 per cent. Samples can become increasingly affected by small amounts of contaminating carbon, which skew the results away from their real age. A sample that is 50,000 years old, for example, when contaminated with 1 per cent modern carbon, gives an age 14,500 years too young. As you can imagine, this is a complete disaster when dating something of the order of 30–50,000 years old, like many of the bones from the archaeological sites mentioned in this book.

Roger came to meet me to see whether there was anything that could be done to improve the situation and obtain more reliable dates. He was particularly interested in dating bone. Bones are attractive for archaeologists, because they are usually either the physical remains of people from the past or they are the remains of animals that they once subsisted upon, so dating bone allows us to date people and when they were living at an archaeological site. At that stage we had just begun using a promising method called 'ultrafiltration' to improve the removal of contaminants from archaeological bones prior to AMS dating.

To date bone, we first need to extract the collagen, the main protein in the bone matrix. Around 80 per cent of bone is hydroxyapatite, which is a carbonate-based mineral. Proteins make up the remaining 20 or so per cent of the bone mass, and 95 per cent of that is collagen. Collagen is a triple-helix shaped molecule, made up of three polypeptide chains each comprising 1,000 amino acids. To extract it for dating we first drill or crush the bone into a powder and then put around 500mg (the volume of about half a teaspoon of sugar) into a test tube, adding weak hydrochloric acid. Slowly the acid dissolves the hydroxyapatite, leaving the collagen. After several cleaning steps we then 'gelatinize' the collagen by heating it in weakly acidic water for twenty-four hours. Slowly the collagen particles untwist from their triple-helix formation and become three single polypeptide chains. This separates the particles from non-collagenous and potentially contaminating particles and allows us to eliminate them. We are left with gelatin, the main component of 'jelly'. Then comes the ultrafiltration.

An ultrafilter is like a sieve, but on a molecular scale. We know the weight of the polypeptide chains of amino acids: each one weighs around 95,000 Daltons.[*] The ultrafilter we use allows us to retain the larger chains, the ones that are over 30,000 Daltons in

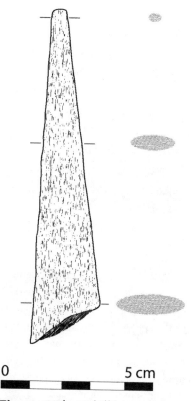

Figure 18 The Uphill bone point.

0   5 cm

---

[*] One Dalton is one-twelfth of the weight of a ^{12}C atom.

size, and to allow the particles smaller than this to pass through the filter. This method allows us to significantly improve the quality of the collagen and to preferentially remove smaller, potentially contaminating forms of carbon.

When I met Roger, we had done some preliminary testing of the ultrafilter method and showed that it did have a significant effect on contaminated bones. We had yet to apply it to actual archaeological cases. We began to do this with Roger. We focused on samples that to him appeared erroneous or at odds with the archaeology. One of these was a sample of a bone point from Uphill, in Somerset. The bone point was characteristic of the Aurignacian, the early modern human stone-tool industry that appears in Europe along with some of our earliest ancestors. The initial AMS date was 28,080 ± 310 BP.*
This English sample was some 4–5,000 years younger than equivalent bone points found in places like France and Belgium. Roger had explained this by suggesting that modern humans moving into the British Isles may have been delayed by deteriorating climate conditions and settled there later than in other parts of Europe.

Roger brought it to Oxford for me to drill into. We carefully took a sample, then extracted collagen and ultrafiltered it. The result was around 4,000 years older, at 32,000 ± 230 BP, perfectly in agreement with examples of the same type of artefact on the continent. It looked as though ultrafiltration was the key step to obtaining a more reliable age.

Encouraged, we extended our work to other sites and eventually

---

* Radiocarbon dates always come with a ± value, which represents the uncertainty associated with the measurement at one standard deviation. In the case of this date there is a 68 per cent chance that the true age will fall within ± 310 years of the age of 28,080 radiocarbon years ago. This radiocarbon date also has to be converted into solar or calendar years to obtain a date in years. This is because radiocarbon is not created in the same amount year by year, it fluctuates, hence the need to calibrate our dates. This is done by measuring samples of known age, usually tree-rings and, beyond ~14,000 years ago, samples of lake sediment, ancient wood, corals and speleothems, the latter two of which are dated using uranium-series methods. The result is a 'calibration curve', the latest of which (IntCal20) spans back to 55,000 years ago and allows us to produce dates in real years.

right across Europe. This gave us the opportunity to work on some of the key French archaeological sites, with deep stratigraphic sequences, spanning the period of the last Neanderthals and the first modern human arrivals. We worked on more than forty sites in Spain, Italy, Germany, France, Belgium and Greece. In 2009, having collaborated with me on several key papers, and become one of my greatest friends, Roger sadly passed away. At just sixty-two, he was at the peak of his powers and in the midst of finding the answers to a lifetime's work. The only thing that kept me motivated to continue was knowing that that is what he would have wanted: to publish the rest of our results and the fruits of almost a decade of labour.

Our principal paper eventually came out in 2014. It had taken more than five years to collect and analyse all of the samples from the many regions we had worked on. We had aimed to explore the temporal relationship between Neanderthals and modern humans in Europe. Our results showed that modern humans and Neanderthals had co-existed for between 2,500 and 5,000 years on that continent, overlapping with one another while not necessarily living side-by-side. We will explore more of the implications of this important work in Chapter 15. The paper was published in *Nature*, with Roger listed posthumously in the last (senior) author position.[2] After that it was time to move on. I wanted to widen my research area and I was very keen to work in Russia.

In 2010 the eyes of the palaeoanthropological world were on Denisova Cave, but there were still major unanswered questions about the Denisova 3 girl and the Denisovan population. One of the big uncertainties at the site was the age of the human remains. There were some radiocarbon dates available, but the results were quite variable in age: some were more than 50,000 years old, others as young as 17,000 years. This confirmed the suspicions of some that there was a degree of mixing of archaeological sediments in the sequence in the East Chamber. In terms of the age of the key Denisova 3 bone, the initial interpretation suggested that it could be as young as 30,000 years old, or more than 50,000,[3] but no one could be sure, due to the mixed nature of the radiocarbon results.

I was convinced that we could do a better job. I thought that if we could date a large number of samples, we could explore the so-called 'taphonomy' of the site and understand whether or not the mixing was a real problem. Taphonomy is a word that describes the science of embedding: how samples of bone, charcoal and so on, came to be deposited in the site. Sometimes material can be 'residual', or already at the site prior to human occupation. Sometimes it can be 'intrusive', or relocated from one part of the site to another, by animal burrowing or human digging and delving. Our job is to choose the best samples and, to avoid the possibility of dating residual material, we usually try to obtain bones that are 'in articulation', which means that they are still in their correct anatomical position. This shows that they have not moved after being deposited in the site. In Denisova Cave, however, these types of bones are non-existent. Instead, we need to select material that is 'humanly modified'. These comprise bones that have tiny visible cut marks, derived from the processing of bones by human flint tools, or artefacts and objects actually made by people. This is the only way that we can date when people were at the site.

The need for more radiocarbon dating work was acute, then, and for it to happen it really required a strong collaboration between Russian and Western scientists. Unlike Europe, where archaeologists had had the benefit of investment in scientific infrastructure and research, Russian colleagues had experienced very difficult times in the post-Soviet years. There were no radiocarbon accelerators in Russia, and although there were many very talented researchers, they were hamstrung by not having the most up-to-date technical assistance. My research was focused on understanding when, and by what mechanism and route, modern humans had expanded out of Africa and into and across Eurasia. While my previous work had focused on Europe, to the exclusion of everything else, I knew that sites in Eurasia had to be brought into focus in order to see the full picture. I was very keen to apply our state-of-the-art dating methods to more sites across the period of the Middle and Upper Palaeolithic transition, 50,000 to 30,000 years ago, to work out when humans, Neanderthals and Denisovans had occupied sites across Eurasia. Anatoly Derevianko was extremely interested in collaborating, and through hard work and good luck I managed to secure the funding mentioned in the previous chapter, which enabled the project to commence.

I went to Novosibirsk in February 2012 to discuss with Michael Shunkov and Anatoly plans for the research project and how it was going to work. It was absolutely freezing. Temperatures average minus 20°C in winter. We focused particularly on the importance of directly dating artefacts and humanly modified bones to be able to securely date human presence at the site.

It's incredible to be able to touch and handle examples of the earliest jewellery and ornaments ever made in Eurasia. Perhaps 45,000 years ago, maybe more, the archaeological record reveals the remains of deliberately perforated animal teeth, of hyena, fox, reindeer and bear. These must have been worn around the neck, either singly or in number, we cannot be sure. Equally, we cannot be sure what purpose they had. Was it simply to look good? Did people enjoy decorating themselves, in the same way we do now, or was there something more? Did this material signify something? Membership of a group, perhaps, or

Figure 19 A perforated reindeer tooth pendant from Denisova Cave.

a shared belief system. I also wonder how it is that these items became part of the archaeological record and found their way into the sediment in the cave. Were they accidentally lost or dropped, or were they deliberately discarded because they were no longer wanted or needed?

Today, these are extraordinarily precious objects, and in order to determine their age we had to propose a very carefully thought-out plan to minimize damage. Over the years of sampling bone, I have developed a so-called 'keyhole' drilling technique that enables us to take samples from precious objects with great care. By making a small hole, 2–3mm wide, in a tooth I am able to use my dental drill internally, widening inside but maintaining only the small external hole out of which comes a steady stream of white dentine or bone powder. Through this method we are able to date tiny samples without significant external damage.

Katerina and I took samples from several trips to Denisova Cave and Novosibirsk back to Oxford for radiocarbon dating. The first set of samples was designed to test the hypothesis that Layers 11.1 to 11.4 in the East Chamber, the resting place of Denisova 3, were extensively mixed in age. At the site, we pored over the cross-section of archaeological sediments. You could see what looked strongly like unmixed sediment, but then in some areas there were patches that appeared to be more mottled and indicative of disturbance. We were aware that over millennia carnivores such as hyenas had been living at the site, producing dens for their cubs in the sediment, and that this could influence the archaeology. Fortunately, with Michael Shunkov

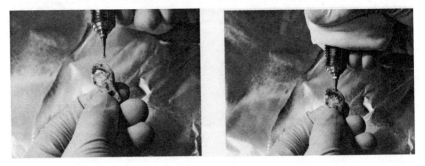

Figure 20 Keyhole sampling of a Denisovan tooth ornament.

and his team, we were able to select reliable samples from what seemed to us to be unmixed levels. We were able to identify some samples that were found centimetres away from the Denisova 3 bone, and others from above and below. Back at the lab Katerina spent several weeks extracting collagen and preparing them before they were ready to run on the accelerator.

It's always exciting when the samples are in the big machine. As noted earlier, it takes only a few minutes to reveal the general ages. I often call the lab to get the technicians to update me on one of the latest batch of samples. I remember asking our AMS technicians to check on the first eleven dates that we obtained. They were sensational. Ten of them were more than 50,000 years old. The one that wasn't was from higher up in the so-called Early Upper Palaeolithic, a period that dates to around 45–35,000 years ago. We knew then that it was highly likely that Denisova 3 was beyond the limit of radiocarbon: more than 50,000 years old.

There was a lot more work to be done. Most of the site's sediments dated beyond radiocarbon and so it was necessary to use other approaches. 'Optical dating' is a technique that allows us to measure the ages of mineral grains exposed to low-level radiation in archaeological sites. The Denisova team had been joined by Richard 'Bert' Roberts and Zenobia Jacobs, two of the best exponents of optical dating and, like Katerina and me, a married couple.[4] They were now working with the Denisova team to put numbers on the layers of sediment that were beyond the range of radiocarbon dating. Crucially, they would be able to date a huge number of single mineral grains from the sediments. Directly dating these allows us to consider whether those grains have moved in the site or not. If many grains of sediment show variation in age this could be due to post-depositional mixing. If they are all the same, it is likely that the sediment is undisturbed.

Sampling for optical dating is hard work. This is because you cannot expose the sediment to any kind of light, otherwise the signal you want to measure will be destroyed and the luminescence clock zeroed. For this reason, it is often done at night. It is not unknown for optical scientists to work until 3 or 4 a.m. in a dark cave taking sediment samples for dating.

By late 2015 we were ready to start analysing the results we had in order to publish them. But there was a major problem. All the dates we had, and there were by now around forty in total, were from the upper parts of the site, while all of the human bones, which everyone wanted to know the ages of, were from below this. As well as being older than the radiocarbon limit, they were so small that they could not be directly dated. How could we estimate their age? Bert and Zenobia would date the sediment in which they were found, but that work would take time.

It was then that Katerina came up with a brilliant idea of using Bayesian modelling to include ages estimated from the mitochondrial mutation rates obtained from each of the seven Denisovan and Neanderthal samples we had from the site. It sounds complicated. Let me explain.

In building the radiocarbon chronology of a site, we commonly use a revolutionary technique called Bayesian chronometric modelling to generate our final results. Thomas Bayes was an eighteenth-century mathematician with an interest in the doctrine of chance and probability. The statistics he devised allow us to include the element of prior or initial belief in our statistical analysis of probability. This is combined with new observations or data to assess the probability of something occurring.

Imagine, for example, that I asked you to assign a probability to the statement 'Tom Higham will go to work on his bicycle'. Initially, without knowing me, you would think the probability is low, because most people travel by car or public transport. You might say it is a 15 per cent chance. But if I told you I live in the centre of a town like Oxford, then your estimate might increase, say to 50 per cent. As more prior information is provided, so your assessment of probability changes. If I then told you that I had injured myself playing tennis yesterday, your assessment of probability might drop back to 5 per cent or even less. This prior information is at the heart of Bayesian inference and statistics. In essence, Bayes's theorem can be summarized by the function: *initial belief × new data α improved belief.*[5]★

Bayesian approaches to radiocarbon dating and chronology were developed in the 1990s and they are attractive because they allow us to include not just the dating results in the analysis, but these other important sources of relevant 'prior' information.[6] Priors in this case usually include the layering of archaeological sediments. We know independently that these are laid down in order from youngest at the top to oldest at the bottom, unless there are exceptional

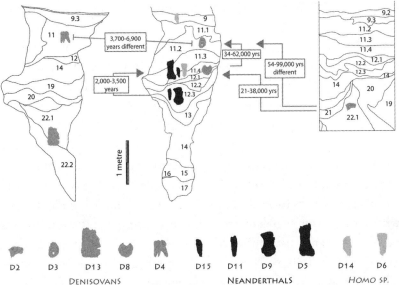

Figure 21 The basis of the Bayesian model we built. The excavated layers in each of the main chambers are shown. The silhouettes show the human remains of Denisovans and Neanderthals, as well as unidentified *Homo*. The lines show the age differences between the fossil bones as calculated using mitochondrial DNA analysis. The difference between Denisova 3 and 4, for example, was between 3,700 and 6,900 years. We used this to order the archaeological sequence relatively, from top to bottom. Radiocarbon and optical ages from the different layers were also used in the model.

★ α means 'is proportional to', so in Bayesian-speak this would be: 'the Prior multiplied by the Likelihood is proportional to the Posterior'.

circumstances like post-depositional mixing. We can also include information from other dating techniques: the ages of coins that were perhaps found in the site, the presence of volcanic ash deposits which are already dated elsewhere at other sites, the fact that a site must date to before European settlement in 1836, and so on. With powerful computer programs, we can combine all of these types of information to calculate new, so-called 'posterior' probability distributions that give us a more precise estimate of the dates of archaeological sites.

Fortunately, my lab in Oxford has one of the world's best Bayesian specialists working in radiocarbon: Christopher Bronk Ramsey. It's rare to use the word 'genius' to describe a colleague, but Chris is just that. Katerina, Chris and I started to construct a Bayesian model for the site with the aim of incorporating the mtDNA mutation rate data along with the other sources of chronometric information we had, to try to determine the age of the Denisovan fossil remains. We were fortunate to be able to use some of Bert and Zenobia's preliminary optical ages to help anchor parts of the model. Initial runs looked exciting, but there was still something that wasn't quite working. The age ranges generated for the fossil remains were quite imprecise and there were worryingly wide uncertainties in our model runs. It turned out that our estimates of uncertainty on the key mutation rates were not realistic. One of the Leipzig genetics PhD students who spotted this problem recommended using a statistical analysis called an Erlang distribution to solve it.* I had never heard of the Erlang distribution, so I went to talk to Chris Bronk Ramsey in the office next door to see whether he knew anything about it.

Of course, he did! After hearing my plaintive question about Erlang-something-uncertainties, he nodded, turned his chair back to face his computer and started to code. In around fifteen minutes, he

---

* The PhD student was Fabrizio Mafessoni. Agner Krarup Erlang was a Danish mathematician, working in the early 1900s. One of his jobs was to work out how many phone calls could be made on the early switchboards and how many operators would be needed to handle them. He created a new branch of statistical analysis in the optimization of networks in order to solve this problem, later named after him. In 1946 the International Telephone Consultative Committee decided to call its unit of traffic intensity an Erlang.

had it all worked out. With a quiet, 'That should work . . .' he pressed the 'Run' button and away the model went. Brilliant!

The model was now much more robust and precise. We finally had the code for a model that would give us numerical age estimates for the Denisova Cave hominins. More than a century after its development, a method devised to calculate the capacity of the world's telephone networks had helped us to date the Denisovans.

The paper was published in *Nature* in January 2019 with much press interest and a big celebration for our group at one of the local Oxford pubs.[7] In the same issue Bert and Zenobia published the results of their complementary work on the optical dating of Denisova Cave.

Our paper showed several interesting things. First, we had dated the ornaments directly and their ages were older than any others we had ever dated in Eurasia – between 43,000 and 49,000 years. This was a very important finding because it might mean, if these were made by modern humans, as we expected, that our human ancestors were in the Altai before most other places in the west of Eurasia. Our dated ornaments were demonstrably earlier than anything similar that had ever been dated in western Europe, for example. Second, we had date estimates for all of the human remains. We dated the earliest

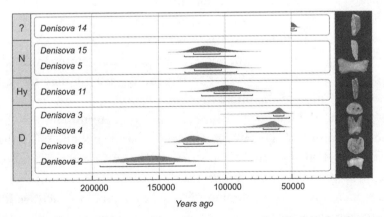

Figure 22 Final ages determined from Denisova Cave for all of the fossil remains. N = Neanderthals, D = Denisovans, Hy is, of course, Denny the hybrid, and the attribution for Denisova 14 is uncertain.

Denisovan, Denisova 2, to as early as 195,000 years ago. Next, we had a group of both Neanderthals and Denisovans around 100,000 to 125,000 years ago. Interestingly, this was around the time of the so-called Last Interstadial, when the climatic conditions were similar to what they are today. This was the date we ascribed to Denny as well. She dates to around 100,000 years ago (79,300–118,100 years). Third, we had a date estimate for Denisova 3, the pinky bone, of 51,600–76,200 years ago. This age estimate overlapped with the date obtained using another genetic technique, based on the nuclear genome, and, not surprisingly, due to the similarities in the mtDNA genome we mentioned in Chapter 6, was nearly identical to the age of Denisova 4, the large tooth in the South Chamber.

Based on the sediment dates and the presence of the first artefacts in the lowest parts of the site, we know Denisova was occupied from perhaps 300,000 years ago. By 200,000 years ago we have Denisovan DNA in the sediments, as well as fossil and archaeological evidence for their presence. At around 190,000 years ago, we also detect and date Neanderthal DNA in the sediments of the East Chamber. The bulk of the Neanderthal fossil remains fall into the warmer Last Interstadial period, around 120,000 years ago. This is the period where we almost certainly have Denisovans present too, to account for Denny, who had a parent of each, of course. Neanderthals disappear from the Altai following this, but, as yet, we are not sure when and why. Denisovans appear to remain. Denisova 3 and 4 are the latest fossil remains we have and these people are at the site as late as 51–55,000 years ago. Sometime after this, Denisovans finally disappear from the cave, never to return. We assume they move to another location or slip into extinction.

The crucial questions that we could not yet determine were when modern humans arrived in the Altai and what they found when they got there. Did they make the ornaments and objects of symbolic importance in the site, and others in the wider region? As yet there is no evidence for modern humans in the key layers in which we find the ornaments; in fact, there are no modern human remains at all. Could these key ornaments have been made by Denisovans? We will tackle these key questions in the next chapter.

10. On the Trail of the Modern Human Diaspora

Tracking the movement of groups of people across time and space in archaeology is difficult. In the Palaeolithic in particular, as we have seen, the presence of human remains is scarce. To track the earliest modern humans as they dispersed across the vast reaches of Eurasia, we have to rely on the careful analysis of stone tools from large numbers of excavated archaeological sites and attempt, where we can, to link human groups to those sites. This is often challenging because for millennia modern humans, Neanderthals and probably also Denisovans made very similar types of tool. Diagnosing a particular maker on the basis of lithic analysis is therefore very difficult. During the period over which the replacement of our archaic cousins by modern humans occurred, however, the so-called Middle to Upper Palaeolithic transition, a range of new tool types appears, and for the first time we are more confidently able to link specific groups or assemblages of artefacts with particular groups of humans. In

Figure 23 Distribution of Initial Upper Palaeolithic (IUP) sites across Eurasia and North Africa. T-P is Tenaghi Philippon.[1]

Chapter 3, for example, you may remember that we mentioned the Aurignacian, a stone-tool industry associated strongly with early modern humans in Europe by virtue of fragmentary modern human remains being found in company with distinctive stone tools of that industry. Similarly, we also mentioned the Châtelperronian, a slightly earlier stone-tool type thought by many to be linked with the physical remains of Neanderthals and therefore made by them.

Some of the most intriguing stone-tool industries of this time are generally termed the Initial Upper Palaeolithic. They fall into the period at the very end of the Mousterian, about 50–40,000 years ago, associated with Neanderthals and just before the Upper Palaeolithic, a period linked strongly with modern humans. I, and many others, have been keen to determine whether we can use this as a signature for the spread of early modern humans across Eurasia. If we can prove the hand of modern humans in this type of stone-tool industry, we have a means of tracking their dispersal across vast tracts of the continent for the first time.

In the dusty Negev desert of southern Israel, on a small hillside by the side of a road, is an archaeological site called Boker Tachtit. If you drive an hour and a half to the north-east you reach the Dead Sea fortress of Masada. A kilometre from Boker Tachtit is the kibbutz of Midreshet Sede Boqer and the retirement home, and graves, of former prime minister David Ben-Gurion and his wife.

In the 1970s, an archaeological team led by Tony Marks came to Boker Tachtit to excavate. They discovered four archaeological layers, one on top of the other, from Layer 1 at the base of the site to Layer 4 at the top. The flint tools they identified were more indicative of a phase of transition when compared with the established Mousterian stone-tool industry of the wider region that had been present for tens of thousands of years.

They painstakingly reconstructed some of the stone tools that were excavated, to discover exactly how they had been made. Flakes of flint previously removed in the Stone Age can sometimes, with perseverance, be slowly fitted back together, like a (very difficult) three-dimensional jigsaw puzzle. This allows archaeologists to reconstruct the so-called *chaîne opératoire*, or operational sequence, and work out the precise methodology of how stone tools were manufactured, step-by-step. It also allows us to see whether pieces of flint that fit back together are found in different archaeological levels, and therefore to determine whether they have moved between layers after being deposited in the site. This can tell us a great deal about whether the material in a site is mixed or has largely stayed in its original burial position. The Boker Tachtit refits confirmed that there was almost no movement of material between the layers, which means that we can interpret the evidence with confidence and track changes through archaeological time.

The people who were living at Boker Tachtit gradually changed the methods they used to make stone-tool points, probably destined to be used on spears and the like. They were mainly using the Levallois method initially, but this changed through time and by the end of the occupation they were using a new bidirectional flaking method to make exquisite spear points.[2] By the final archaeological layer, Layer 4, the Levallois method had essentially disappeared and the stone tools were of the Upper Palaeolithic type.[3] Levels 1 to 3, on the other hand, were Emiran: an Initial Upper Palaeolithic industry named after another site where similar tools were first found. These levels were clearly part of the transition away from the Mousterian Levallois-based stone toolmaking tradition that had held sway in the region for thousands of years.

There is evidence that these changes could all have happened quite quickly. There is a joke in Israel that you can see the transition to the Upper Palaeolithic happening on a single stone tool. The few

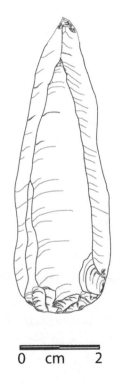

0   cm   2

Figure 24 An Emiran point (redrawn after Belfer-Cohen and Goring-Morris 2017; see note 2).

radiocarbon dates from Boker Tachtit suggest that this transition from Middle to Upper Palaeolithic took place as early as 50,000 years ago. New excavations have been conducted over the last few years,* specifically to obtain new chronometric evidence, but this is not published at the time of writing, so the early dates cannot yet be verified.

So, who was making these Initial Upper Palaeolithic tools? Was it Neanderthals, who we know were in the region very early, or could it have been early modern humans? In the search for ancestral dispersals, archaeologists constantly look for evidence of something new, something novel, that could be the result of a population movement into a new area. Could this stone-tool industry be part of it?

The Mediterranean strip of Israel, Palestine and Lebanon is a key region in these discussions of our late human evolution. We have evidence for modern humans here over 120,000 years ago, perhaps even earlier, at the sites of Skhul and Qafzeh, which we mentioned in Chapter 2. Then something interesting happens. Modern humans disappear and are replaced by Neanderthals, around 60–70,000 years ago. This has been, for many decades, the only place we have evidence for this type of replacement. Usually, where we see replacement, it is by supposedly superior modern humans of the assumed inferior Neanderthals. In the Levant it seems to be the opposite.

Marks and his colleagues compared the Emiran stone-tool industry they found with other sites in the wider Levant region. The most important of these was the site of Ksar Akil, which is just north of Beirut.

Ksar Akil was first excavated by three Jesuit priests (Fathers Ewing, Doherty and Murphy) between 1937–8 and 1947–8, and then later in the 1970s by a French team. The archaeological sequence covered the entire period from the Mousterian all the way up to the last 10,000 years. It is an astonishing 23m deep and sits just under a huge rock overhang. In levels XX–XXIII, two-thirds of the way to the bottom of the site, the team had excavated a stone-tool industry that Marks could see was identical to material from some of their Boker

* Led by the Israeli archaeologist Omri Barzilai.

Tachtit levels.[4] What makes the site of great importance, however, is what the Jesuit priests found there.

In 1938 the diggers came upon the bones of two humans. The better preserved of them was the skeleton of a child, about eight years old. It was buried with a protective covering of water-worn boulders. They christened the find 'Egbert'. Fragments of a second person were found close by. Excited by the discovery, but chilled by the imminent approach of the Second World War, the Jesuits decided to halt the excavation. They covered the burial in four inches of concrete to protect it until they could later return. Instead of taking the shortest route back to the United States, Father Ewing decided to make his return eastwards, ostensibly to avoid the deteriorating situation in Europe. He travelled via Baghdad, Basra and Bombay, eventually making it to the Philippines, where, absorbed and distracted by anthropological study of the tribes on Mindanao Island, he stayed from 1939 to 1940. Sadly, this was to prove a grave mistake, for while there he was captured by the Japanese army and imprisoned for three long years ('*internity*' as he called it).[5] The precious excavation

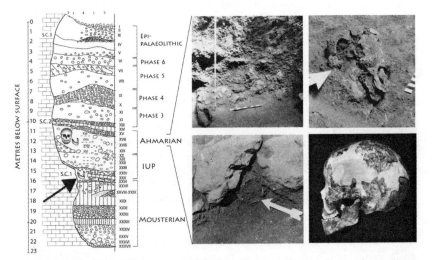

Figure 25 The stratigraphic section (left) shows the succession of archaeological layers in the site of Ksar Akil. The location of Egbert is shown (pictured right during excavation and after reconstruction). Ethelruda is shown left, arrowed in its position in the IUP.

notes, films and photographs he had with him from this first period of excavation were seized and, sadly, destroyed.

After the war, in 1947, he and Father Doherty returned to the Lebanon determined to finish the job they had started, and renewed excavations until 1948. They dug all the way to the bottom of the site. Egbert's remains, encased in a tonne of rock, were removed and taken back by ship to the US. It took Ewing a year to painstakingly chisel out the skull.[6] In 1953, in triumph, it was returned to the Beirut National Museum. During the chaos of the civil war in Lebanon from 1975, though, it was lost and has never been seen again. The casts made by Ewing are all we have. Fortunately, on the basis of these it is possible to identify Egbert as a modern human.[7]

Vast numbers of excavated animal bones from the site were sent to London's Natural History Museum to be analysed before being shipped on to Leiden's Rijksmuseum van Natuurlijke Historie, which later became the Naturalis Museum. In the late 1950s a partial fragment of a human jaw was noticed lying amongst the other bone material. It was later christened 'Ethelruda'.* It too has been identified as a modern human. Through a careful analysis of the box in which it was found it was possible to work out exactly where the bone came from: Level XXV. Ewing had written that this level was 'definitely associated with an important change in geology and lithic tradition'. We now know it was the start of the Initial Upper Palaeolithic phase of the site. In the same layer, near to where Ethelruda lay entombed, there was a single Emiran point, the classic stone-tool point representing the industry. It looks, then, as though modern humans were responsible for making the Initial Upper Palaeolithic at Ksar Akil.

My wife Katerina Douka has worked extensively on the Ksar Akil site, in particular the chronology of the sequence. In 2015 we finally got to Beirut and went to Ksar Akil with Dr Corine Yazbeck of the Lebanese University. The site is very different now from the time of the Jesuit excavations. Development on all sides of the steep limestone valley in which it lies has changed the surrounding scenery and

---

* Ethelruda was thought to be lost, rather like Egbert, but she was recently found in storage boxes in the National Museum in Beirut.

threatened the preservation of the site. Although it is officially protected, industrial buildings have encroached very close. It is still possible to see the dim outline of the old excavation square in the grass and to imagine the hot dusty work that those Jesuit priests did excavating and sorting the tonnes of material from the site. Thanks to them we can be more confident that the Initial Upper Palaeolithic of the Ksar Akil and probably Boker Tachtit is likely to be the work of modern humans.

Archaeologists have drawn together a range of data from other sites that have similar stone-tool industries, and the distribution of these sites is striking. We find the Initial Upper Palaeolithic in North Africa at the site of Haua Fteah, in Europe at sites like Brno-Bohunice in the Czech Republic, in the Near East, and across a swathe of Central Asian territory. The most easterly sites spread all the way to the Altai, Lake Baikal and even to China. Similar industries, but hugely dispersed geographically.[8]

There seem to be at least three possible explanations for this pattern. First, it might indicate the movement of an idea across a landscape, as different groups copy one another. Second, it could be the result of the dispersal of a group of people from one place to another, carrying the same or a similar cultural tradition. Third, it might be a result of convergence, in which different groups, isolated from one another, independently develop similar technologies.

To really answer the question, we will need to explore the archaeological record associated with the appearance of the Initial Upper Palaeolithic in more detail, using modern scientific methods and bioarchaeological approaches. We will need to confirm whether the same human group was responsible for it in all regions, or whether, rather like the Mousterian, there was more than one maker.

I have been involved in research over the last decade on a number of these Initial Upper Palaeolithic sites, from France to North Africa, Georgia to Russia, trying to do just this. One of the most intriguing and important sites I have worked on is the Grotte Mandrin, in the Rhône Valley of southern Mediterranean France.

Mandrin sits atop a rocky crag some 225m above the eastern bank of the Rhône. The Rhône is a significant river, second only to the Nile in terms of the deluge of freshwater it discharges into the Mediterranean 120km to the south of the site. Today the valley is part of an arterial route through France from north to south, and it was doubtless the same in prehistoric times.

Colleagues have been excavating here for more than fifteen years.* I travelled to the site for the first time in 2007. I confess, I was not blown away by anything except the view on that first occasion. It seemed a small and not hugely significant site. It turned out, however, that it was far richer than I thought: more than 60,000 stone tools and over 70,000 bone remains have been excavated in the years since. Amongst this proliferation of material was something very significant. It was a stone-tool industry, identified previously at a tiny number of sites in the region of the Ardèche (a *département* in southeastern France that lies between Lyon and Montpellier), that was unlike anything seen before. It wasn't Mousterian, that was clear, being characterized by a different range of tools, from blades to tiny bladelets, all made on different types of stone than the Mousterian. It had been called 'Néronian', after another site nearby where similar material had been found in the 1960s, but the industry was so different from anything else that my colleagues were convinced it had the hallmarks of an incoming group. In their view it must have been made by modern humans. What was intriguing was that after the Néronian ended at Mandrin, the Mousterian had returned, so here we had what is termed an 'interstratification' event, in which Mousterian Neanderthals had been replaced at the site by moderns, who had then later been replaced once more by Neanderthals. Which is a remarkable thought.

And there is other very interesting evidence at the site. The opening of the shelter faces directly north, into the teeth of the Mistral winds that blow cool and moist air from the north. The Mistral blows for around a hundred days a year on average, speeding up as it is funnelled through the Rhône Valley on its journey south. It averages

---

* My colleagues Ludovic Slimak and Laure Metz are the principal excavators of the site.

50kph but can gust up to 100kph at times. Once, on the top of Mount Ventoux, its winds were measured at 350kph.

In Level E, three tonnes of rocks were found that appeared to have been arranged deliberately into a circle. The centre was practically clean of material. The purpose of the rocks seems clear: whoever was living there at the time had carefully constructed a windbreak, a shelter to protect themselves from the gusting Mistral winds. One can imagine wooden stakes and skins augmenting the stone structure rather like a Native American teepee. If this interpretation is correct, this would be one of the earliest pieces of evidence of a formal built shelter in human history, some 30,000 years or more before we find anything else like it.

When the team looked for the closest parallels to the Néronian stone-tool industry in other sites they could see that the material from Ksar Akil's Initial Upper Palaeolithic levels XX–XXV and at sites like Boker Tachtit was the closest. Full analysis of the stone tools revealed that the Néronian industry contains evidence for projectiles that appear to be different from anything else seen before in Europe. This has been suggested to include the earliest evidence in the world for bow-and-arrow technology. If true, this is astonishing.

I have worked for over a decade with the excavation team on dating the site, and particularly the key level, Level E, that contains this Néronian industry. Surprisingly, every single date we obtained was just beyond the reach of radiocarbon, that means greater than 50,000 years ago. I thought this meant that it could not be modern human, since it was just too old for their presence in Europe. We therefore used luminescence methods and more radiocarbon dates from higher up the sequence and eventually managed to show that the Néronian dated between 49,000 and 53,000 years ago.

Could this really be modern humans? If so, it was earlier by far than anything we had ever seen before in Europe.

Then something was found that sealed it.

Human teeth are found throughout the archaeological sequence. In the Mousterian levels, all of the teeth are Neanderthal, as we would expect in western Europe. In the two main archaeological levels following the Néronian layer they are Neanderthal too. In the

Aurignacian levels, around 41,000 years ago, modern human remains are present, as expected based on material found at other sites with this industry. In the Néronian level, they found a single tooth. When its morphometrics and shape were plotted against other hominins, including Neanderthals and modern humans, it sat clearly with the modern humans. It was a modern human tooth, 8,000 years earlier than any other evidence in Europe.

At Mandrin, then, it looks as though we have something completely new. Instead of Neanderthals being replaced by modern humans in Europe around 41–43,000 years ago, it seems that moderns entered significantly earlier. They did not persist, however, and were themselves replaced by Neanderthals. It was only around 7–8,000 years later that modern humans once again repopulated the area, this time in greater numbers and at many more sites.

This is the first time we have found an interstratification like this in Europe: first we have Neanderthals, then modern humans, then Neanderthals again, then modern humans. This is obviously extremely significant, but how can we explain what happened at Mandrin? How could humans have entered Neanderthal territory so much earlier than we considered possible? I have two thoughts on this. The first concerns the population of Neanderthals in Europe 50,000 years ago. We have evidence that they were under some degree of stress in terms of their genetic diversity, which was low (I will discuss more of this in Chapter 15). A study of mitochondrial DNA in thirteen Neanderthals showed that they appeared to have gone through a bottleneck and a population contraction.[9] We can imagine Neanderthal groups perhaps isolated from one another around 50,000 years ago, and present in lower numbers than before. Perhaps this was the moment when modern humans made this initial incursion. But why did moderns not remain in Europe? Why did Neanderthals replace them later on?

I think that one hint is found in the climate records.

In north-eastern Greece there is a large area of former peatland called Tenaghi Philippon that provides crucial evidence. Tenaghi Philippon has yielded one of the longest and most detailed records of climate change in Europe: a 200-metre-deep core, spanning 1.35 million years, tells us about the changing climate that has characterized the region. The core shows that 54,000 years ago the Earth's climate was entering a phase of interstadial or warmer conditions. As we have seen previously, the Earth's climate in the Ice Age was subject to a cyclicity such that every 1,000 years or so it flipped from very cold to warm and then slowly moved back to cold again. At 54,000 years ago the climate was switched to warm. This period, known as Interstadial 14, lasted for much longer than the usual millennium, however; in fact, it was the longest phase of interstadial conditions that we have had in the last 70,000 years. As the climate warmed, so trees in these mid-latitudes responded. Tenaghi Philippon records a surge in pollen from temperate deciduous trees like oak, away from species typical of the cold steppe. The key ingredient that enabled this at this time was increased precipitation. Further to the south, records show an increase in the freshwater deposited by the Nile into the Mediterranean in response to increased humidity and milder climates in North Africa. This was caused by the cycle of monsoonal rains moving to the north of the continent. It may have influenced environments that were favourable to the movement of modern humans out of Africa and into the Levant, and perhaps from there into wider Eurasia. The dates of some of these early Initial Upper Palaeolithic sites match this scenario.

At 48,000 years ago, however, the climate switched again in a way that was particularly bad for humans.[10] This was H5, a so-called Heinrich event that followed a collapse in the margins of the gigantic Laurentide ice sheet that covered most of northern North America. In prehistoric times, icebergs periodically calved in vast numbers and entered the Atlantic. This huge influx of frozen freshwater had the potential to affect the passage of warm salty seawater across the Atlantic to the margins of Europe. When that source of warmth was switched off, temperatures would plummet to Ice Age conditions in Europe and the climate deteriorate dramatically. The age of the H5 event coincides with the end of *Homo sapiens* occupation at Mandrin. They either died out or left. People did return to the site later, but those who did were Neanderthals. H5 might be the key to understanding what happened at Mandrin and why humans ultimately failed then to remain in Europe. Perhaps they were not as well adapted

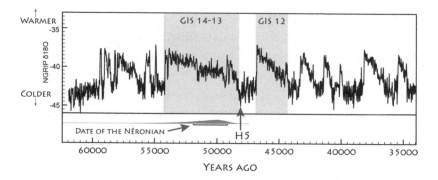

Figure 26 Comparison of the date of the Néronian modern human occupation at Mandrin with the H5 climate event about 48,000 years ago. It would be more than 4,000 years until modern humans were next seen in Europe. GIS stands for 'Greenland interstadial'; GIS-14–13 is the longest phase of warmer conditions we see during this long period of time. H5 – the Heinrich event arrowed, brings this to a shuddering halt.

to the extreme challenge of cold conditions and could not cope with it. One can almost imagine the desperate way that the people living at Mandrin tried to cling on. They built structures to protect themselves and their children from the increasingly cold and difficult conditions. But it was not enough. Their relatively brief presence in France ended and Neanderthal domination of Europe resumed shortly afterwards. Neanderthals succeeded in Ice Age Europe 48,000 years ago, while modern humans went locally extinct. Thousands of years were to pass before Neanderthals would encounter modern humans again in this part of the world.

Meanwhile, far to the east we still had unfinished business. Our dating work at Denisova had focused on determining the age of the human remains, but we had also generated important data on the age of the many ornaments in the site. As you will recall from the previous chapter, these had been dated to between 43,000 and 49,000 years ago, but we still did not have an answer to the more complex and difficult question of who had made them. The majority of specialists would link these objects with modern humans, because in Europe they are almost exclusively linked with us. However, it is possible that Neanderthals were also producing pierced tooth pendants not dissimilar to the Denisova examples, as we saw in Chapter 3.

Could the Denisovans have been responsible for producing similar ornaments far to the east? Our Russian colleagues certainly thought so. They based their conclusion on a careful analysis of the stone tools in Denisova Cave, from the bottom to the top layers.[11] According to them, there seems to be a slow and steady development of the tools in the Denisova sequence from the Middle Palaeolithic to the Initial Upper Palaeolithic, demonstrating *continuity* rather than an arrival of this technology from somewhere else.

In November 2017 I was in Novosibirsk with Sam Brown to take more samples for our ZooMS work. After a late evening of drinks with our translator (which comprised rather too many cocktails in the colours of the Russian flag that were ignited prior to drinking), a

Figure 27 Meeting with (from left to right) Maxim Kozlikin, Michael Shunkov and Anatoly Derevianko in Novosibirsk. Not feeling particularly great at this point . . .

surprise meeting had been scheduled at 10 a.m. the following day
with our Russian colleagues Anatoly Derevianko and Michael Shunkov. They wanted to go over the draft dating paper in detail,
especially regarding aspects of modern human presence in the Altai.
A very good idea, but perhaps not at that hour of the day, or in our
hungover state. For more than five hours in Anatoly's very warm
office we discussed all the most detailed aspects of the paper and their
and our interpretations of it.

Both Michael and Anatoly pointed out that there were no remains
of anatomically modern humans at Denisova during this period, or
in the Altai region. The most parsimonious view, they said, was that
Denisovans were responsible for the Upper Palaeolithic and the ornaments. My view at the time was different. I thought it was more
likely to be modern humans, and I felt we had some strong arguments in our favour (although I was prepared to believe anything
anyone said if they would only let me go home and lie down for a
sleep). First, the physical evidence of Denisovans predated 50,000
years ago. Denisova 3, the youngest Denisovan specimen, dated to
just before the date range of our ornament results, as we saw in the
previous chapter. Second, there was the Ust'-Ishim human . . .

In 2011 a *Homo sapiens* bone was sent to my laboratory for radiocarbon dating. I didn't know it at the time, but it had been found in a
remote part of Siberia near a village called Ust'-Ishim by a man who
was hunting for ancient mammoth tusks.* These items, frozen since the
Ice Age, can fetch huge amounts of money, up to US$30,000 each in
some cases, it is said. He spotted a strange bone poking out of an eroding
bank of the Irtysh river near Ust'-Ishim. Recognizing it as likely to be
human, he showed it to a forensic scientist friend who saw its possible
significance and it eventually found its way to Palaeolithic researchers.

When we analysed the sample, a large femur, we could see virtually
no radiocarbon, suggesting it must be very old. To our great surprise
the result we calculated ranged up to 47,000 years old. To be sure, we
repeated the dating work three times and each time it was the same age.

* His name was Nikolai Peristov.

As we worked in Oxford, in Leipzig Svante Pääbo's team was extracting DNA from the same specimen. They found that amongst the expected modern human DNA sequence, there was something else that was very surprising. They identified quite large chunks of Neanderthal DNA dotted throughout the genome. The DNA fragments were longer than those we see in the genomes of modern people, indicating that the Ust'-Ishim Man had lived closer to an ancestral introgression event with Neanderthals. As we get closer to the date of introgression, the stronger the non-*Homo sapiens* genetic signal is, and the easier it is to accurately date the admixture. Imagine

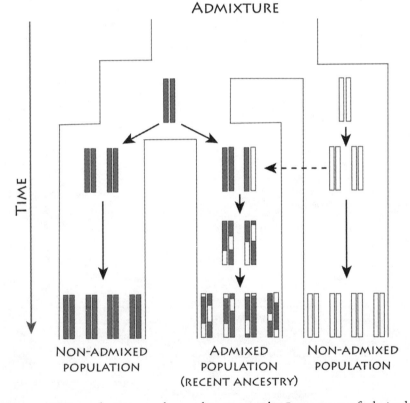

Figure 28 How admixture and recombination works. Large tracts of admixed DNA are evident closer to the introgression event when two previously separate populations interbreed. The larger the block sizes, the closer to the interbreeding event (denoted by the dashed arrow) you are.

if a Neanderthal and a modern human had interbred, they would produce a child with an average 50 per cent split of Neanderthal and modern human DNA. If that child had interbred with another modern human, the DNA from the Neanderthal would be reduced again by about half. Gradually, with successive generations in which only modern humans are left to breed with, that ancestral Neanderthal DNA would be cut into smaller and smaller chunks. As we saw in Chapter 5, this is called 'recombination' and is how we can sometimes calculate how far back introgression occurred on a generational timescale.[12]

We calculated that the flow of Neanderthal DNA into humans occurred between 232 and 430 generations prior to the time the Ust'-Ishim Man lived. That would mean that sometime around 55,000 years ago the ancestors of Ust'-Ishim had met and interbred with

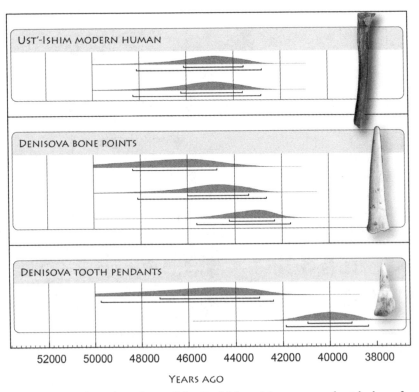

Figure 29 Radiocarbon dates of the Ust'-Ishim Man compared with dates for bone points and ornaments from Denisova Cave.

Neanderthals.[13] This gives us an earliest age for the introgression of DNA from Neanderthals to us.★

Our date on Ust'-Ishim put modern humans on the ground in Siberia by 45–50,000 years ago, right in the middle of the dates obtained for the ornaments from Denisova. When we compared the ornament ages with the dates for Ust'-Ishim there was a significant overlap. To me this suggested it was just as likely that *Homo sapiens* had made these objects as that Denisovans had. Ultimately, however, we could not prove it either way. We needed more evidence before we could be sure.

As fate would have it, we did have something that could shed further light on the problem. In July 2016 we had struck it lucky in our ZooMS work at Denisova Cave. We had found not just one but two new hominin bone fragments, from the East Chamber. Denisova 15 came from Layer 12 and Denisova 14 was found in Layer 9.3. Layer 9.3 is an Early Upper Palaeolithic archaeological layer that we had been focusing on precisely to address the question of who made that industry. Both bones were tiny, just over 20mm in length, but there was enough weight of bone for dating, isotopes and, of course, DNA.

Figure 30 The two tiny bones found with ZooMS: Denisova 14 on the left and Denisova 15 on the right. Two angles for each bone are shown. These two tiny fragments were the first human bones we had found since Denny, after screening a total of more than 3,500 bone fragments.

★ In the wake of the discovery of Ust'-Ishim, Anatoly Derevianko sent out teams to scour the banks of the Irtysh river, to see whether any more skeletal remains or archaeological evidence could be found. They combed over 300km of riverbank, but found nothing.

The radiocarbon date we obtained for Denisova 14 was 46,300 ± 2,600 BP – again, putting it directly alongside the ornament dates as well as Ust'-Ishim – and an age fitting closely with that estimated for the Initial Upper Palaeolithic in the cave. If this turned out to be a modern human, it would provide additional evidence to support their being the makers of the Initial Upper Palaeolithic.

Once again, we had to wait patiently for Viviane Slon and the others in Leipzig to work their genetic magic and tell us what we had found. Just under a year later Viviane had the answer from our tiny Denisova ZooMS bones. Would Denisova 14 be a modern human as we thought? Or would a tentative link between Denisovans and the Initial Upper Palaeolithic in the east of Eurasia be forged?

First, Viviane had done mtDNA extraction and sequencing. Ninety-nine DNA sequences were extracted from Denisova 14, of which fifty-nine looked to be consistent with hominin DNA and forty from animal DNA. If that sounds encouraging in terms of hominin reads, it isn't: by contrast, Denisova 15 yielded 15,804 identifiable sequences that were all hominin mtDNA. Denisova 15 had mtDNA of the Neanderthal type.

Viviane next explored what kind of DNA was present in Denisova 14 by comparing it against reference mammalian mtDNA databases. It turned out that 2.1 per cent of the identifiable sequences were human. The DNA fragments, however, were not ancient, because they had none of the chemical uracil modifications, those C to T ratios that characterize ancient DNA. They all looked to be contaminating DNA segments from modern sources. Other fragments, which were ancient, were only from mammals: hyena, horse, deer and others. Damn it all! We had to conclude that no ancient hominin mtDNA was preserved in Denisova 14. The only DNA was from other taxa and modern contamination.

It was a huge disappointment. After finishing her Masters dissertation, Sam had been working for weeks preparing samples for ZooMS for our project. Since Denny, number DC1227, she had sampled more than 2,000 bone fragments before we found Denisova 14 and 15, so it was a long time between ZooMS hits. Although we had a probable new Neanderthal in the form of Denisova 15, Denisova 14 was really a potential key to unlocking the answer to the question of who made the Initial Upper Palaeolithic at Denisova. Sadly, it just wasn't to be.

Our Russian colleagues believe that Denisovans made the Initial Upper Palaeolithic tools and wore the ornaments that have been found there. Perhaps they are right. Perhaps I am underestimating the abilities of these people, culturally and cognitively, just as we did with the Neanderthals before them. It may well be that the situation is not simple and there was more than one maker of these tools and ornaments. At the sites we have looked at in Europe and the Levant, it appears so far that modern humans were involved and the transition to the Initial Upper Palaeolithic proceeded very much as a *Homo sapiens*-linked process. During my career I have learnt that the road one must travel to work in the Palaeolithic comprises five or six major bumps followed by one stunning insight and then some more bumps. More work, patience and data would be needed to firm up our tentative conclusions at this stage.

It turned out, however, that we had one final shot at working out who made the ornaments at Denisova and other sites in the east of Eurasia, and it came from a very unexpected source: dirt.

11. DNA from Dirt

One Friday evening in Oxford in spring 2001, I was in a pub with some of the geneticists from the Ancient Biomolecules Centre when we were joined by someone I'd never seen before. The visitor was a freshly arrived Danish post-doctoral researcher. He looked like the bass player from a heavy metal band. He wore cowboy boots and had long hair and a very gruff voice. I would later discover that, as a teenager, he had got lost in the snowy wastes of Siberia and suffered frostbite (on a testicle no less), and had survived only by making a fire and keeping it lit all night. The name of this adventurous Viking spirit was Eske Willerslev, and he would later become one of the world leaders in ancient DNA.

In 1999 Eske had been the first to recover ancient DNA from glacial ice. He wanted to see whether, in the cold regions of the world, and perhaps other places, it might be possible to extract DNA from soils and sediments. I was tasked with helping him to radiocarbon date some of the sites he was working on while in Oxford and later.

To me, Eske's idea about DNA in sediments seemed somewhat fanciful. What if DNA moved downwards with percolating water and moisture? Surely such a method would work only in places where this could be shown not to happen, such as permafrost regions. Wouldn't DNA just dissolve away in soils, subject to post-depositional processes of degradation? And what about contamination from things like animal faeces and human agriculture?

Eske and his team subsequently established that, in fact, it was possible to extract and sequence DNA from sediments and identify it as clearly ancient. The most spectacular example was material that he obtained from the very bottom of Greenland's ice sheets, down over 2,000m. At the base of one of the deep Greenland ice cores, at more than 400,000 years old, he and his team found DNA from plants like spruce, alder, pine, birch and a range of herbs commonly found in open boreal forests.[1] They also found DNA from beetles and butter-flies. Eske was able to show that at one time Greenland had a completely different environment from that we know today.

In 2003, in Siberia, his group found nineteen different types of plant DNA in sediment cores. In New Zealand they retrieved DNA from the extinct giant moa bird in sediments in caves.[2] Eske said that if genetic signals of plants and animals could be routinely extracted from sediments in archaeological sites like this, then it would have major implications. One could take a core down through a site, for example, and link successive occupation layers with different human groups, based on their DNA. There would be no need to take destructive samples from human bones and one could avoid the issues of modern human contamination in ancient human bone samples.

At the time this seemed to be more like science fiction than science to me, but Eske's comments were to prove extremely prescient.

Other researchers were also exploring cave sediments to see whether human DNA evidence might be preserved in archaeological sites. A dry cave in Arizona that had amazing levels of preservation in some of the organic remains inside appeared the ideal place to test the method. Disappointingly, the results were not great.[3] Only very short sequences of DNA could be extracted, and these turned out to be highly contaminated. Dating the sediment and the DNA within it was also hugely challenging because of the potential movement of both in archaeological sites.

In Oxford, other sediment DNA work confirmed the general pessimistic view I held. While Siberian permafrost DNA seemed to produce reliable results, work at a couple of sites in New Zealand showed that, in pre-European layers, sheep DNA was present.[4] This suggested that DNA had leached down through the sediments into the pre-European layers when sheep clearly were not yet present in the islands. The prognosis for DNA in the kind of caves and sediments where humans had once lived seemed to be somewhere between 'next to impossible' and 'probably never'. But, as we have seen throughout this book, in the world of genetics, the impossible has a tendency to become possible. Over the last decade, sediment

DNA analysis, and the recovery of human DNA, has started to become a reality.

Matthias Meyer, a researcher at the Max Planck Institute in Leipzig, started experimenting with extracting DNA from sediment in archaeological sites in 2014. Might DNA from Neanderthals and other long-dead humans be extractable? New developments in methods to exclude modern human contamination and improvements in isolating demonstrably ancient DNA meant that the time was right to revisit sediment DNA.

In 2015 I found myself with Svante Pääbo and Viviane Slon in Tautavel, near Perpignan in the south of France, visiting Professor Henry de Lumley, the doyen of French Palaeolithic archaeology, and his wife Marie-Antoinette to see whether we could secure access to their collection of human fossil remains for dating and genetics. It was a tough trip because the de Lumleys were understandably resistant to the idea of us drilling holes in their precious collection of ancient fossils. It had taken a considerable amount of work simply to organize the meeting. Fortunately, one of my post-doctoral scientists, Thibaut Devièse, had worked at Henry's site of Caune de l'Arago in Tautavel, and had a degree of inside knowledge. He helped us to arrange a meeting. We met Svante and Viviane at Toulouse airport and drove south to the Mediterranean coast.

The de Lumleys' collection of human remains is in a strongroom at the centre of their institute in the beautiful French countryside, in locked metal cases that few are privileged to see. Like conjurors, with white gloves, they slowly take out their precious fossils, each in an individual box, and reveal the contents to you, describing how they were found and pointing out interesting features. It is literally science as performance art.

Over the more than fifty years that they have been excavating here they have found almost 150 human bones from thirty individuals, but their greatest discovery was Arago XXI – the almost complete face of what is generally attested as a 460,000-year-old *Homo erectus* individual. Henry and Marie-Antoinette's great pride in finding these lost human remains is obvious. Each fossil marks a point in the diary of their lives' work and brings memories flooding back of people, dates, even times of the day. The more you see the love in their eyes, the more you know how difficult the moment will be when you propose drilling a small amount of material from the tooth of this or the ear bone of that, for ancient DNA or dating.

They have so far resisted every attempt to take samples from Arago XXI, and in response to our sampling question, Henry suggested that before tackling the human material we might want to test the recovery of DNA on some of the less valuable animal bones from the site. We agreed not to touch the human material this time, but to focus on only the animal bones. However, when Svante asked whether we could perhaps take samples of soil sediment from the Caune de l'Arago to see whether DNA could be retrieved from there, they were much, much more accommodating. After all, dirt is just dirt, right? Thus it was that later in the day Viviane and Svante were able to take several small samples of sediment from the cave and bring them back to their laboratory for analysis along with the animal fossil remains.

I was sceptical about DNA preservation at the site; it seemed just too hot and dry, and the site much too old; but of course it was worth testing it to see. Sadly, it turned out that I was right; no DNA at all was recoverable from the sediments there.

But there were other sites being explored. Six, in fact, including Denisova Cave. Perhaps there'd be better luck elsewhere.

Perhaps I ought to explain here just how DNA can be preserved in cave and other sediment sites. A lot is to do with how DNA binds chemically with minerals in sediment, particularly with clays and silts. Temperature also plays a role, in the same way that we have seen with bone preservation. Why some sediments in sites preserve DNA and others do not is an area under investigation. To explore whether DNA can leach downwards it is necessary to check each archaeological site and test whether DNA from species not likely to be present are identifiable, or whether sediments that are 'sterile', or have no archaeology in them, contain DNA. It is also crucial to ensure that these sterile layers are sampled to the same degree as the archaeological levels in order to check the integrity of the sediment for DNA movement. If the sterile layers have no DNA, that's good; if they have DNA that has percolated down from higher up, that's obviously bad.

While Arago had failed to yield any DNA, it emerged that at the other sites across Eurasia with Pleistocene sediments that had been sampled, the situation was very different. The developments that lay behind NGS or Next Generation Sequencing, including the identification of DNA that was damaged and therefore ancient, were enabling the recovery of ancient DNA from sediment. The word was that the Leipzig team were getting genuinely ancient DNA from Palaeolithic archaeological sites.

In the initial sampling work, they had obtained sediment from sites ranging from 14,000 to 550,000 years old. Sixty-one out of the eighty-five samples they had analysed contained DNA that was demonstrably ancient and could be aligned against existing genetic sequences from different animals. They had identified twelve different types of mammal in the sediment samples, including hyena, horse, mammoth, bovid, wolf and more. The identifications harmonized with the types of animals previously identified at the sites by archaeologists.[5]

Incredibly, nine of the samples actually contained demonstrably ancient human DNA. From eight of those nine, the DNA extracted overlapped with parts of the Neanderthal genome in around 95 per cent of the cases. For the first time, extinct human DNA had been found in sediments, and not from one site but from four.

At the Belgian site of Trou Al'Wesse the Leipzig team found Neanderthal DNA in a site where no human fossils had ever been discovered before. At El Sidrón in Spain they found only Neanderthal DNA in a site where there are only Neanderthal bone remains, those, as you may recall from Chapter 3, of thirteen or so humans who were killed and their remains deposited in a chamber below with no associated animal bones.

The ninth sample of sediment with ancient human DNA disclosed something different. In 84 per cent of the sequences extracted, there was overlap with the Denisovan genome. That sample came from Layer 15 in the East Chamber at Denisova Cave. Another first.

The DNA fragments from the Denisova Cave sample containing

Neanderthal DNA were found almost certainly to come from one person. One of the sub-samples yielded a huge amount of DNA, probably due to a small, unidentified chip of Neanderthal bone in the sediment.

How does the amount of hominin mtDNA in fragments of bone compare with that found in sediment? Between 28 and 9,142 DNA fragments per milligram were extracted from bones at the site, while in sediment it was between 34 and 4,490 mammalian mtDNA fragments.[6] Surprisingly, then, large quantities of DNA can actually survive intact within cave sediments. This raises the possibility that we can detect the presence of humans at sites where no skeletal remains have ever been found.

It seems that we don't even need bones now to do archaeology; just sediment will do in some cases. It is incredible but true to say that if we had never found any physical remains of Denisovans, no bones or teeth, we would still know about them now from the sediment DNA work alone. It's a truly remarkable feat of science that we can do this work.

Inspired by the sediment DNA work and the observations of the Leipzig team regarding human DNA, I wanted to test whether there was human DNA recoverable from some of the mysterious Initial Upper Palaeolithic sites of Siberia. I wanted to see whether we could link this industry here to early modern humans, rather like in western Eurasia, or to Denisovans and Neanderthals. I had my eye on several key sites in the Transbaikal region.

In the summer of 2018 I headed east to Ulan-Ude, the capital of the Republic of Buryatia, about 100km south-east of Lake Baikal and the home of Russian Buddhism. Ulan-Ude was closed to foreigners until 1991, and its tiny airport belies the fact that more than 400,000 people live in the city, whose main attraction is a 38 tonne, 7.7-metre-high head of Lenin, reputedly the biggest in Russia.

Our plane landed at 4 a.m. and on arrival at my hotel I was told that it was full and anyway my booking was (mistakenly) made for the following night. Fortunately, the upmarket Baikal Plaza hotel was just around the corner. The first thing I saw in the lobby there was a picture of Vladimir Putin, who stays here when visiting and who has had a special fish dish in the hotel restaurant named after him. As I gazed out of the window at Lenin's massive head I realized that this was clearly the place to be in Ulan-Ude.

The following morning, I headed south-eastwards with colleagues in a couple of four-wheel drives. First stop was the archaeological site of Varvarina Gora, where Russian archaeologists had previously uncovered Initial Upper Palaeolithic archaeology. We worked to sample sediments systematically across one of the trenches previously left, until darkness stopped work. That night we stayed in a nearby village called Novaya Bryan. It is dominated by a large factory that made truck parts, disused since Glasnost and Perestroika destroyed the local economy. Now it's used occasionally as a set for dystopic Russian movies.

Near Novaya Bryan is another site, called Kammenka, where evidence had been found for Initial Upper Palaeolithic archaeology that included ornaments and a putative bone whistle, along with a pit packed with animal bones. It was thought that this might be some kind of provisioning store for difficult times 45,000 years ago when human ancestors were living there. We spent a long hot day excavating and sampling sediment from two of the layers in the site.

Later our convoy drove further north-east, across rolling countryside and steppe grassland to the site of Khotyk. On a couple of occasions our journey was halted at a crossing by the passing trains of the Trans-Siberian Railway. Khotyk rests in the hug of a small hill overlooking the wide valley of the Ona river. In the early 1990s a sequence of three Upper Palaeolithic levels was excavated, the lowest of which, dating to around 41,000 years ago, contained a fragment of a flute made from a swan bone. Incredibly, our human ancestors, most likely I think to be *Homo sapiens*, played music at this ancient time. In Germany, eight similar flutes in a better state of preservation, made from mammoth ivory, swan and griffon vulture bones, have been found in Aurignacian layers.[7] One, a vulture-bone flute from the southern German site of Hohle Fels, is 22cm long and has five preserved finger holes and two carved V-shaped notches, probably to blow into.[8] It was recovered in pieces and reconstructed painstakingly from a dozen fragments of worked bone over two years. The archaeologists who excavated them think that flutes, and possibly percussion instruments, were an important part of everyday life, because they were found alongside evidence of daily activities and not kept in any kind of special place. The Hohner company reconstructed one of the flutes, and they play similar notes to modern recorders. I was once at a conference in Germany when one of these, made of a swan bone, was played. It was beautifully haunting and evocative. Our ancestors were not brutes living in caves; they made art and music, just as we do today. Their inspiration was the natural world, the world in which they lived and survived. We can imagine that items like this might have also played a role in maintaining links and wider networks with other groups, bringing people together on occasion to socialize or celebrate or mark the passing of time, and helping to enrich their world.

The Khotyk flute is a small fragment, 4–5cm long, but unmistakably the same type of instrument as the German equivalents far to the west. I was able to examine and take a tiny sample from it for dating in 2017. It is a truly wonderful feeling to pick up and touch part of a flute that was once lifted to the lips of a person who, 40,000 years ago, played music under the wide skies and expanse of the Russian steppe.

At the bottom of the Khotyk site was a dark layer around 5cm thick, which contained evidence for Middle Palaeolithic archaeology. We carefully sampled the sediment along a one-metre line spanning the layer. I imagined that here we might find human DNA evidence for Denisovans or perhaps even Neanderthals. Either would be incredible and, if found, would extend the known range of the distribution of both groups far to the east. We could only hope that the preservation was good enough. As evening fell, I counted the samples collected: 132 small plastic bags of sediment from three key sites.

We returned to Denisova Cave two days later and met Viviane and Svante there, when I gave them the samples to transport back to Germany. Despite not being superstitious, I kissed the bag containing the samples before handing them over, and hoped for good luck. We had worked very hard to get them.

Nine months later, in April 2019, the results were in. Using a small set of samples from each site the Leipzig team had targeted mammalian mtDNA to see whether ancient DNA was preserved. For those that gave positive samples, they went on to specifically target hominin mtDNA.

At Varvarina Gora they had initially screened eight samples of the twenty-nine we had obtained, targeting mammalian mtDNA fragments in them. Unfortunately, all of them were negative.

At Kammenka, an initial batch of ten samples were tested from the site and two were positive for ancient DNA preservation, but in both cases the signal was weak, with a few sequences from horse and rhino. Of forty-two samples in total then tested from the site there were no traces of ancient hominin DNA in any of them. This was disappointing, because ancient DNA did seem to be preserved in the sediment to some extent.

There was more encouraging news from Khotyk. Fourteen samples from the site were initially tested, and twelve were positive for ancient mammalian mtDNA. Species such as horse, rhino, cow and more were identified, although, in a few cases, the signal was weak.

Then came the news we had hoped for: there were detectable traces of ancient hominin DNA, but in just two of the forty-seven additional samples that were tested from Khotyk. Both samples came from the Level 3 Initial Upper Palaeolithic horizon.

The first sample yielded insufficient data to determine a hominin group of origin. The second sample, however, contained mtDNA that appeared to be of the modern human type. Viviane, who had been leading the DNA lab work, warned us that this was a tentative assignment based on a few DNA fragments, but this was hugely encouraging news. Our first sampling trip to Siberia had actually yielded human DNA from sediment. This was exciting, not simply because we had identified DNA but because the sites we had looked at are open-air ones, not caves, and therefore the preservation should not be as good.

The fact that it seemed likely to be modern human mtDNA was stunning, but the news was tempered by the fact that there had been no DNA recoverable from Level 4, the Middle Palaeolithic. We had not been able to demonstrate a Neanderthal or Denisovan presence this time. This would have to wait. Perhaps another site or another research team would identify this.

Meanwhile, while we worked in the Transbaikal, more sediment

DNA had been analysed from Denisova Cave.[9] Of the 728 samples analysed, a whopping 22 per cent of them had hominin DNA in them. Amongst these samples were stronger signals of modern human DNA in the sediments containing Initial Upper Palaeolithic archaeology.

Gradually, through the evidence from 'dirt DNA', it is beginning to look as though the makers of this industry in this part of Siberia were modern humans, as they were in Lebanon at Ksar Akil and Grotte Mandrin in France. Recognizing that this is preliminary and more work is needed, though, I have to curb my enthusiasm. We also have to remember, as I said earlier, that linking a single stone-tool industry to one hominin type is difficult, particularly when we also have so much emerging evidence for interbreeding. It's possible, for example, that a largely hybrid group could be responsible in some sites or that Denisovans or Neanderthals could have made similar types of stone tools in another. Still, this is the beginning. With more effort and a bit of luck I feel confident that we can soon solve some of these archaeological problems that have seemed so intractable and out of reach for so long.

This science is so new that, at the time of writing, the earlier-noted paper is the only one to have been published with actual data. Since then, the Leipzig team has worked on more than a hundred archaeo-logical sites and analysed almost a thousand samples of sediment for DNA, including from throughout Denisova Cave itself. Around one sample in ten has provided ancient DNA. The lab has built auto-mated systems to allow the work to take place more rapidly than ever before. Robots now work to process the samples in the more mun-dane parts of the preparation and they are getting through a huge number. Sediment DNA has the potential to unlock so much about our past, and more of the truth about what happened in the Middle and Upper Palaeolithic when modern humans met Denisovans and all of the other characters in the prehistoric Middle Earth.

Who knows what other, as yet unidentified hominins, might be lurking in the dirt?

12. The Hobbits

In 2004 a new hominin *was* discovered. Technically it was found 'in the dirt', but not in the vestigial DNA remaining in sediment. Actual physical remains of this one were found in a cave on the island of Flores in Indonesia. Like the Denisovans later, this created huge interest around the world and provided the first hints for palaeoanthropologists that there was much more complexity in the human family than we had ever appreciated.

Henry Gee is an editor at the leading scientific journal *Nature*, and

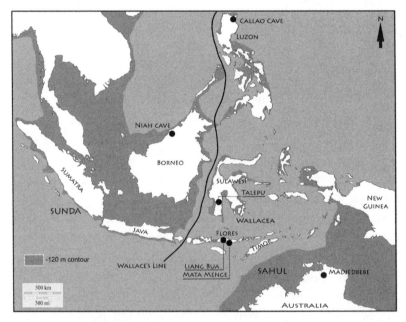

Figure 31 Sites and locations featured in this chapter. The sea levels from 30,000 to 80,000 years ago fluctuated between 50m and 120m below current levels. In this illustration the −120m contour is shown to give an idea of the maximum size of the additional landmass of Sunda and Sahul. Wallace's Line is that modified by T. H. Huxley.

he regularly receives new papers for review.* When asked which paper, of the more than 15,000 he has received in his thirty-odd years there, was the most jaw-dropping and impactful, his response is immediate: 'the Hobbit paper'. Absolutely no hesitation. The day that the two *Nature* papers on the Hobbits of Flores appeared in print, 28 October 2004, was a landmark date – a day when the human family became bigger, though in the case of the Hobbits, perhaps we should say a little bit bigger.

Certainly, it brought us closer to confirmation that the Earth was richer in human diversity than had been previously thought. Indeed, the discoveries of new human groups are still happening, the latest as recently as January 2019, and we now know there were at least five different groups of humans living at broadly the same time 40–120,000 years ago, and probably for several tens of thousands of years before that. The 2004 discovery sparked an increased and continuing level of interest in the islands of Southeast Asia.

Island Southeast Asia was a very different place prior to 10,000 years ago. Many of the islands that we are now so familiar with did not exist. During the Ice Age huge tracts of now drowned land were exposed due to the vast volume of water stored in frozen ice towards the poles. Sea levels fluctuated between 30m and 120m below what they are today.[1] The sea level was so low that Australia became part of a much larger continent that we call Sahul. Sahul comprised Australia and New Guinea and all the land between and surrounding the two. Around the time of Australian settlement by humans, thought to be 50–60,000 years ago, the sea level was 60–75m lower than today. It was possible to walk from New Guinea to Tasmania. Although Sahul was never connected to the mainland of Eurasia, in Island Southeast Asia a similarly exposed vast land mass called Sunda existed. Sunda included now drowned land connecting the island of Borneo to peninsular Malaysia and the Indonesian islands of Sumatra

---

* Only around 7 per cent of the scientific papers submitted to *Nature* are accepted for publication. A whopping 60 per cent of submitted papers are rejected without even being reviewed.

and Java. Dispersing groups of hominins would have been able to move freely across Sunda too, but to get to Australia it would have been necessary for our human ancestors to 'island hop' across stretches of open water up to 100km wide. Only in the Holocene period (the last 10,000 years on Earth) did sea levels begin to rise and then stabilize. The current sea level was reached only in the last few thousand years.

Between Sunda and Sahul lay the region of 'Wallacea', named after the nineteenth-century evolutionary biologist Alfred Russel Wallace, the man who had dead-heated with Darwin in devising the theory of natural selection. He had noticed differences in the endemic fauna of the Asian continent to the west, dominated by placental mammals, and the Papuan/Australian to the east, dominated by marsupials. He drew a line between the two denoting the biogeographical boundary, now known as Wallace's Line.* Only a tiny number of terrestrial mammals have ever crossed Wallace's Line in prehistory: humans, rodents and Proboscideans (*Stegodon* or dwarf elephants).[2]† A major question in palaeoanthropology is which kind of humans crossed and when.

Mike Morwood is the archaeologist who must be credited with making the Hobbit story happen. His initial archaeological work was mostly in northern Australia, exploring the likely pathways for the arrival of the first modern humans onto that continent. He began to be drawn to the question of where and how the initial pulse of dispersal from Sunda to Sahul had occurred. So he decided to work in Indonesia, and the island of Flores struck him as having interesting potential.

Flores sounds just like something from *Peter Pan*, and resembles a science-fiction story in terms of its natural history. Dwarf elephants (*Stegodon*), giant lizards (Komodo dragons, currently only

---

* Wallace's Line was modified by T. H. Huxley, a contemporary of Wallace, who included the islands of the Philippines east of the line (with the exception of Palawan); it is his modification that is illustrated in the figure at the start of this chapter.
† One might add rhinoceros to this list if one accepts that, with Huxley's revision, the Philippines ought to be considered east of Wallace's Line. See note 25, for example.

extant on the eponymous island, of course), giant tortoises, giant storks, crocodiles, frogs, bats and monkeys have all lived on Flores at one time or another. Rather like in the movie *The Princess Bride*, there were also rodents of unusual size, including rats the size of small dogs. They are still there today, called *betu* in the local Manggarai language. We know that several species of these rats have been present on the island for hundreds of thousands of years. Island isolation, caused by the waxing and waning of sea levels in the case of Flores, often provides curious evolutionary pressures upon endemic fauna; examples of gigantism and dwarfing are often observed as natural selection runs free and creates new solutions to the puzzles of life. In 1964 the so-called 'island rule' was outlined by zoologist J. Bristol Foster, who invoked it to explain the influences of island life on phenotype.[3] This rule was based on a range of previous observations on islands that showed that some types of animals become smaller over time, particularly those that formerly had large bodies, whereas smaller ancestral animals sometimes become much larger. Flores is a classic example of this variability in shape and size.

I have worked on dwarf mammoths from Wrangel Island in Siberia, sampled dwarf hippopotamus from Greek Mediterranean islands and many giant flightless birds from New Zealand and Madagascar, but the Flores fauna is like nothing else.

Could humans isolated on islands also be subject to these kinds of evolutionary pressures?

Morwood decided to focus initially on a site previously excavated by Dutch archaeologists, called Mata Menge. There, they had found human-made artefacts more than 500,000 years old, characterized by stone implements made of local volcanic rock.[4] This old age suggested that *Homo erectus* had probably been on Flores. This fascinating discovery meant that this group must have had some kind of ability to cross maritime barriers, since according to sea-level reconstructions, Flores was never less than 60km from the mainland.

In the early 2000s the team moved to another site, called Liang Bua (which means 'cold cave'). This is a very large and sunlit cave in the centre of Flores, 500m above sea level. It has a very deep sequence of archaeological levels of over 10m and perhaps, thought Morwood, up to 17m. It seemed that this might be the right place to hunt for evidence for the first modern human arrivals on Flores and, because of the possibility of finding very old sediments towards the bottom of the cave, to link it with the story emerging from Mata Menge.

Liang Bua was an attractive location for humans to live in prehistory: it has a water source nearby as well as high-quality stone for toolmaking. Its large interior space had been used for a school in the late 1940s shortly before the Dutch amateur archaeologist Father Theodorus Verhoeven started excavations there.* The site had also been extensively excavated previously by Indonesian archaeologists, principally by Raden Panji Soejono, who had led excavations in the cave from 1978 to 1989. In 2001, Morwood's excavations began, with Indonesian colleagues led by Thomas Sutikna and a team of local and international researchers as well as Soejono.

In 2003, towards the end of another productive season of digging, Morwood and the Australian contingent were due to head back home, leaving the work for another week or so to the Indonesian core team only. A few days later, excavating at a depth of almost six metres in Sector VII, a 2m × 2m excavation square in an ancient Pleistocene layer of the site, the trowel of Benjamin Tarus, one of the team, hit something in the clay sediment.

Benjamin's trowel had sliced through part of the left brow ridge of a small human skull, judging by its size probably from a child of around 5–6 years.[5] At a great depth, in a Pleistocene layer, near stone tools and *Stegodon* bones, they had found human remains.

Morwood had arranged to call Thomas Sutikna every evening for updates of the day's work on site. Nothing prepared him for the call he made on 10 August 2003. Morwood said that Thomas sounded as though he had been sitting over the phone for an age, waiting for it to ring so he could blurt out the news. Morwood was ecstatic; his gamble had paid off.

The excavators had one big problem, however. The bone material

---

* Verhoeven excavated an initial test dig near the west wall in 1950 and undertook a bigger excavation in 1965.

was soft, with a texture rather like papier mâché. Removing the human remains without distorting or altering their shape was going to be a major challenge. It was going to take days to dry the bones after they had been exposed prior to removal. Somehow, they had to speed this up as well as conserving the precious material as it was being exposed in the excavation. After some quick thinking and experimentation, and within the limitations of the resources available on the remote island, they decided that a mix of pure acetone, nail-polish remover and UHU glue would be the best method for preserving the finds. By preserving the bones as they were excavated, and removing them in blocks, they found they were able to do the best job of extracting the precious human cargo. They turned their hotel room into a makeshift conservation lab and bought up almost the complete stock of nail-polish remover on Flores. The painstaking job of preservation continued steadily, night after night, for two solid weeks.

In due course it emerged that they had found not only a skull, but also the bones of the same skeleton's right leg, connected in anatomical position with the tibia and fibula, as well as the kneecap, a pelvis, most of the foot bones, parts of the back bone, shoulder and collar bones, fingers, toes, as well as ribs.[6] Only the arm bones were absent. The skeleton would become known as 'LB1' – Liang Bua 1 – and it would prove to be one of the most important human skeletons ever found in palaeoanthropology.

It was small; no more than one metre tall. When the team were finally able to begin the process of cleaning up some of the finds, however, they were in for an almighty shock. An examination of the teeth in the jaw revealed that the teeth were all erupted and worn. Despite the very small size of the human and the initial thought that it was a child, this could only mean one thing: the skeleton they'd found was actually of an adult. An exceptionally small adult.*

---

* Thomas Sutikna told me about the shock of seeing all the teeth in a complete state, but with some of them worn. He showed his team mate Rokus Awe Due, who was more experienced in faunal and human bones, and silently the incredible implications of their discovery dawned on them: 'There is no "word" expressed from both of us,' he recalled. 'Instead, I began to draw the skull in tracing paper and sent it via fax to Prof Soejono and Mike.'

In due course the human remains were carefully transported to Jakarta. Morwood made plans to return immediately to Indonesia, and arranged for his colleague Peter Brown, a physical anthropologist, to come as well, to look at the new finds and identify just what they had got on their hands.

It took around six seconds for Brown to see that the tiny skull and postcranial bones that had been painstakingly excavated and conserved were of a completely new human species. It was definitely *not* a modern human. Initially, however, he kept his opinions to himself. This was important, because dark clouds were gathering in the background. Raden Soejono was determined that the physical remains should be passed to Professor Teuku Jacob, the grand old man of Indonesian physical anthropology, rather than Brown. Jacob was Soejono's best friend and had been given all of the remains from his initial set of excavations, which were mostly from the Neolithic. None of Jacob's findings on these had ever been formally published, and many researchers had found it impossible to study or even view their excavated remains while they were under Jacob's control. Morwood was determined not to allow this to happen; the precious fossils must not be permitted to disappear into storage at Jacob's institution.

Tense negotiations followed. Eventually, after a great deal of back and forth, it was decided that Brown would analyse the material with the aim of producing publications on both the human material and the site excavation.

For a week Brown and Sutikna carefully cleaned the skull and postcranial bones, with Morwood hovering and offering nervous support. Brown removed sediment from within the LB1 skullcap in order to measure its cranial capacity. Mustard seeds were poured through the hole at the base of the skull, the so-called foramen magnum,* to estimate its cranial volume (he had smuggled the seeds

---

* Meaning 'great hole' in Latin; this is the basal hole in the skull through which pass the spinal cord and major arteries and which tells palaeoanthropologists a great deal about posture and bipedality. Mustard seeds are used as a standard of sorts, since they are of a uniform size and spherical in shape. Lead shot was formerly a favoured material but it is not suited for travel due to its weight and effects

through airport customs to do this). It was tiny, measuring only 380cc (by comparison, modern people have brain sizes that average 1,400cc). Brown repeated the measurement three times; the result was always the same – the brain size was about the same as hominins that lived 2–3 million years ago. For Brown, this meant only one thing: LB1 could *not* be a member of the wider human family of *Homo*; it was simply too small. He was expecting something of the order of 600cc, which would make it close to the size other palaeoanthropologists would consider the minimum for a member of the *Homo* line. Because the specimen was smaller than this, Brown felt that it simply could not be human, so he suggested the new species should be named *Sundanthropus tegakensis* ('erect apeman from the Sunda area').[7]

Morwood was not convinced by *tegakensis*, however, arguing that it was only local Malays and Indonesians who would recognize the name. He suggested using *floresianus* instead, in honour of Flores. Fortunately, one of the referees of the subsequent paper pointed out that *floresianus* means 'flowery anus',[8] so in the end this was changed to *floresiensis*. Of greater importance, however, was the genus name: was it to be *Sundanthropus* or perhaps *Homo*?

During the subsequent review of the papers, referees suggested that the physical anthropology meant that the specimen fitted better into the human family and therefore should be *Homo*. Morwood agreed that it appeared to be more like a small human than a primitive early human ancestor. Although its brain size was closer to Australopithecines such as the 3.18-million-year-old Lucy (*A. afarensis*) specimen, its brain endocast indicated a similar suite of features to *Homo erectus*, with enlarged frontal lobes, just on a far smaller scale. Ultimately, it was determined that if *Homo habilis* could be called *Homo* by Richard Leakey and his team in Kenya, then there was no reason why LB1 was not to be classed as of the same genus.

---

on a fragile fossil. To measure the cranial capacity the mustard seeds are poured in and the fossil gently shaken, followed by more pouring and shaking until the entire cranium is full. The seeds are then poured into measuring cylinders to determine the volume. This is usually repeated several times to determine the confidence of the measurement (Brown, pers. comm.).

As we will see, although complex, this argument appears to be justified in the light of subsequent work.

The team members decided that it would be useful to create a nickname for these new humans and, of course, it is now widely known that they were termed 'Hobbits'. Initially, Brown was sceptical about this, but it stuck in the wake of the massive publicity that the papers generated.

The press covered the story worldwide. It was huge. I well remember the iconic image of the Hobbit that was published at the time, of a male hunter with a spear and a large rat slung over his shoulder.* A few months after the papers were published, the University of Oxford Museum of Natural History obtained a cast of the LB1 skull. I went to visit one day and held the tiny reproduction in my hands, marvelling at the size of this latest member of the human family: one metre tall! It was, without doubt, one of the most sensational finds in the field of human evolution in probably more than a century.

Not everyone agreed. In the wake of the press coverage of the publications reporting the new finds, Teuku Jacob, and other researchers in Australia, including Maciej Henneberg and Alan Thorne, had reached very different conclusions about the Hobbit and what it was. They argued that it was not a new hominin at all, but rather its stature and small appearance were due to a pathological condition. They argued that LB1 was a microcephalic modern human.[9] Alan Thorne said disdainfully: 'There are a lot of people in my field who cannot recognize a village idiot when they see one.'[10]

Their interpretations were rebutted strongly at the time and even more strongly later after new evidence emerged in 2005, when more human remains were found in Liang Bua.[11] Another jaw confirmed the morphology of the first: small and lacking a chin. An upper limb provided key information on the relative limb-length ratios to allow scientists to understand better the body shape and

---

* It is often pointed out that the LB1 skeleton was of a female and the image was wrong. Richard 'Bert' Roberts, in whose office building at the University of Wollongong in Australia the original painting hangs, told me that the artist, Peter Schouten, was informed that the skeleton was probably of a female, but that he painted it as a man for reasons of artistic licence.

proportion. Prior to this it was assumed to be human-like, just small. In fact, a much more primitive postcranial anatomy was implied. Hobbits appear to have had short legs and large feet. In addition, the wrist was very different and more ape-like than other members of the genus *Homo*.[12] It seemed far-fetched to expect that all of these remains could derive from a group of identically diseased individuals. It was far more likely that we were looking at a genuinely new hominin group.

Another significant setback for those suggesting a pathological explanation for the Hobbits came with the publication of new dating work. One of the most surprising aspects of the initial reporting of the *Homo floresiensis* discovery had been the young age inferred for the human remains. Radiocarbon dates obtained from the same depth as the Hobbit bones were as recent as 18,000 years ago, contributing to the idea that this might have been the latest date for their survival. This, in turn, implied that they must have overlapped with early modern humans, who were probably present in the wider region 45–50,000 years ago, or even earlier, making it more likely, in the view of Jacob, Henneberg and others, that the Hobbits were not separate and new but simply diseased modern humans.

In 2016, however, the initial age estimates were shown to be completely wrong.[13] Using new luminescence-based approaches, researchers established that the true age of the skeletal remains and the deposits within which they were found ranged from 60,000 to 100,000 years ago, while the stone tools in the site dated to between 50,000 and 190,000 years ago. The problem with the initial dates revolved around the presence of a 'pedestal' of remnant sedimentary deposits containing the Hobbit remains, situated near sloping deposits comprising much younger sediments from which the dates in the original paper were obtained. Usually we expect archaeological layers to be broadly horizontal, but cave sites like Liang Bua can be more complicated and the movement of sediment from reworking by water or other agents can often occur. In the case of Liang Bua, new excavations between 2007 and 2014 revealed a much greater complexity in the sedimentary sequence and confirmed that Hobbits were present at the site well before the arrival of modern humans.

This suggested that the stone tools in the site were almost certainly made by the Hobbits, not *Homo sapiens*.

How, then, did the Hobbits come to be, and what is their position in the wider human family tree? And why are they so tiny compared with us and other hominins?

Palaeoanthropologists have attempted to determine the ancestry of *Homo floresiensis* by comparative analysis of skull, tooth and post-cranial shape with other hominins. Some aspects of the cranial morphology of LB1 echo *Homo erectus*, for example the cranial vault is broad relative to its height, like erectines. A detailed analysis of the teeth suggests strongly that, while producing a unique dental arcade not seen before in humans, *Homo floresiensis* shared several dental characteristics with Early Pleistocene *Homo* but none with substantially earlier *Homo habilis* or Australopithecines.[14] This suggested that the Hobbits were more likely to derive from an earlier population of *Homo erectus* that probably became dwarfed, once isolated on Flores. But if this did happen, when, and how?

In 2014, at the Mata Menge site, crucial new data were obtained that suggest a considerable antiquity to these humans of Flores and shed further light on their evolution. The team excavating there found six teeth and a mandible fragment that are very similar in shape and size to the Liang Bua *Homo floresiensis* remains. They are, perhaps, even smaller, although one has to entertain the possibility that they simply reflect the intra-population variation in size, so we should be careful not to overinterpret this.[15] The age of these fossils was determined at around 700,000 years. In addition, work at the nearby site of Wolo Sege indicated that hominins there had reached Flores by 1 million years ago.[16] An analysis of the stone-tool evidence from these sites suggests a similar technology characterizes these earliest examples and the later Liang Bua tools that are firmly associated with *Homo floresiensis*. If the Hobbits' ancestors had reached Flores around the time of this earliest million-year-old archaeological evidence, then this would suggest that they had reached their diminutive size at some point over the 300,000 years of time that had elapsed between the occupation of Wolo Sege and the time of the Mata Menge hominin remains.

There seem two possibilities for these ancestors of the Hobbits. First, they arrived on Flores already small in size. Second, based on the island rule mentioned earlier, they arrived large and then became smaller through time. In terms of the first scenario, the teeth of the Hobbits being more akin to *Homo erectus* than to earlier and much smaller hominins, such as *Homo habilis* and the Australopithecines, it is important to also note that these latter two hominins are significantly earlier than the Hobbits at *c.*1.6–2 million years or older, and have only been identified in Africa thus far. *Homo erectus*, on the other hand, is much more widely found and has been identified in East Asia, China and Indonesia, as well as in Africa. This hominin was clearly expansive and left Africa after *c.*1.6 million years ago. On this evidence alone it appears most parsimonious to conclude that it was *Homo erectus* that is the ancestor of the Hobbits – without completely eliminating the possibility that a small pre-*Homo erectus* hominin may have left Africa before about 2 million years ago and given rise to Hobbits on Flores.

One area of uncertainty regarding this hypothesis, however, is whether it is even possible. The problem is that the reduction in mammal body size due to insular dwarfing is usually linked with a more moderate brain size reduction in comparative terms. Insular dwarfing does not simply result in an equivalent reduction in size in all parts of the body; there are scaling factors involved in the reduction of the brain size. The challenge is working out what that is. We have never witnessed island dwarfing in humans so we have to rely on comparison with other animal species. It is clear that, if Hobbits are derived from *Homo erectus*, it is a serious challenge to explain the extent of the reduction in cranial size. The cranial size of *Homo erectus* ranges between *c.*700 and *c.*1,200cc, whereas for Hobbits, as we have seen, it is 380cc. The reduction in size required for insular dwarfing to explain this is far larger than anything that has been seen in other animals. It is very difficult to obtain a brain the size of the Hobbits' by applying the scaling factors known in other mammals found dwarfed on islands.

In 2013, however, a new study of the LB1 skull using micro-CT scanning showed that the Hobbits' cranial capacity was *larger* than previously thought. Instead of 380cc it was actually 426cc; the original estimate was affected by the presence of previously undetected small pockets of unremoved sediment.[17] This changes the mathematics of scaling somewhat, but still not quite enough. If we consider *Homo erectus*'s average of 860cc, for example, and the scaling of the braincase with respect to body size from one to the other, *Homo floresiensis* would end up with a brain size of between 522 and 585cc: still much too large for the actual Hobbits.

Other research has indicated that the mechanics of insular dwarfing could be more complex. Studies of Madagascan hippopotamuses show that their dwarfed brains are more than 30 per cent smaller than they ought to be when compared with their continental ancestor, even after accounting for allometric scaling. Selective pressures on brain size that are independent of those operating on body size may also have played a role in the shrinking cranial capacities we see in Hobbits. Brains are expensive and it may be that in an environment of low calorific availability it is adaptively advantageous to have a smaller brain.

A strict interpretation of models of brain-size reduction under conditions of island dwarfing, then, shows that it is more likely that the ancestor was a much smaller hominin, such as *Homo habilis*. In addition, larger statistical comparisons of the physical anthropology of Hobbits and other hominins have suggested that *Homo floresiensis* is more closely related to *Homo habilis*.[18] The problem is that we have no evidence for this group outside Africa and, therefore, if this is true then it means there was a hitherto unknown movement of a pre-*Homo erectus* hominin out of Africa and into the Old World. Given the unexpected twists and turns that we have already seen in our story of the wider human family, this would not come as a tremendous surprise to me.

Another alternative might be that earlier Southeast Asian *Homo erectus* were smaller than we think, or it could be that there is a founder effect operating, as in cases where a new population is established from a small and uncommon sub-group of a population.[19] Work and debate continue amongst researchers. Geneticists from the Max Planck in Leipzig, for example, have been searching for DNA from the

Hobbits and sediment in Liang Bua that could tell us more about their ancestral origins. So far they have been unsuccessful; the DNA does not preserve well in the wet tropical environment.

Sadly, Mike Morwood will not be one of those continuing the search for our human origins in this most fascinating region. He became ill with cancer and died in Darwin, en route back to Indonesia, in 2013, less than a decade after his pioneering research to discover the Hobbit.[20] He is buried in Wollongong, the site of his last university posting, on a sunny coastal slope looking eastwards out over the vast Pacific Ocean.

In 2007, in a cave on Luzon island in the Philippines, there was another twist in the story of humanity. Filipino archaeologist Armand Mijares and his team were excavating at a site called Callao Cave. Mijares had only returned to the cave to recommence previous excavations in the wake of the Hobbit discoveries, below 26,000-year-old sediments that had already been excavated, to see whether more ancient evidence could be found.* At almost three metres below the cave floor the team found the remains of a human third metatarsal, one of the bones of the foot. It was broken in two. Dating of bones nearby the find spot indicated an age of roughly 67,000 years ago.[21] An initial comparison of the bone with other hominins implied that the bone found its closest relatives in the genus *Homo*, particularly the smaller-bodied versions of *Homo* such as *Homo habilis*. Could this be the crucial piece of evidence supporting a link to this much earlier hominin?

More excavations in 2007, 2011 and 2015 revealed more human remains: two bones from the hand, seven teeth and a femoral shaft (the thigh bone), representing a minimum of three people – two adults and a child.[22] Like *Homo floresiensis* before it, the remains were a confusing mixture of features, some extremely ancient and others

---

* One of the legacies of the work of Mike Morwood was an increased tendency in excavations in the region to dig deeper, in case more material such as Hobbits were found lower down.

reminiscent of *Homo sapiens*. The molars were very small, with the patterns on their cusps echoing *Homo sapiens*. The foot bone was virtually identical in 3D space to those of 2–3.7-million-year-old hominins such as *Australopithecus afarensis* and *A. africanus*.[23] The relative size ratios of the teeth are most similar to *Homo erectus*. Both toe and finger bones have a curvature usually seen in hominins that habitually climb. Like the Hobbits it is extremely difficult to find links to other potentially ancestral humans. The remarkable novelty of the human remains warranted a new species designation, *Homo luzonensis*.[24] Another new human species had been found in Island Southeast Asia.

As in Flores, there is also evidence of a much earlier hominin presence on Luzon. In 2018 researchers announced the discovery of a butchered rhinoceros alongside stone tools at more than 700,000 years ago.[25] Luzon was never connected via a land bridge to the Asian continent. How did these hominins reach the island? It is difficult to be sure, but some researchers suggest drift or accidental voyage, perhaps on a natural raft following a tsunami. Prevailing thought is that maritime technology was a much later development, associated only with modern humans.[26] There are strong currents that flow broadly from the north to the south through Wallacea. If drift voyaging occurred, therefore, it is more likely that the people were coming from the north, rather than the west. If this is true, then we cannot exclude Taiwan or the coast of China as the source of these early pioneers.[27] I wonder, however, whether we might again be underestimating the skills of ancient hominins. Perhaps there was a greater degree of deliberation and technical skill earlier in the Pleistocene than we assume.

These fabulous discoveries in Island Southeast Asia have provided further evidence that the Earth was a primevally complicated place 50–100,000 years ago, a Middle Earth in which the human family comprised a multitude of different types and sizes, living in a variety of ecosystems. This variety occurred owing to the isolation of populations from one another, which set them on a divergent evolutionary path. The islands of Southeast Asia provide fertile ground for evolution to experiment and run wild. Given the number of islands in the region, one must surely expect that other members of the human family are yet to be discovered here. My bet for the next major finding is the island of Sulawesi. Excavations by an Australian team at a site called Talepu between 2007 and 2012 revealed stone tools around 120–194,000 years old in association with the bones of extinct animals, such as *Stegodon*.[28] Could this be the material signature left behind by another new hominin, or could it be related to the presence of Denisovans or *Homo erectus*? Only new excavations and the discovery of fossil hominin remains will confirm this, and for that we will have to wait.

13.   The Journey to the East of Wallace's Line

The genome of the Denisova 3 girl provided the revelation that living people in New Guinea and Bougainville in the Solomon Islands had obtained some of their ancestry from Denisovans.[1] Quite how it was that people living so far from Denisova Cave had Denisovan DNA in their genomes has remained something of a mystery. Some researchers have suggested that the reason might be that Denisovans were living east of Wallace's Line.[2] In this chapter, I want to investigate this conundrum, follow the trail of the

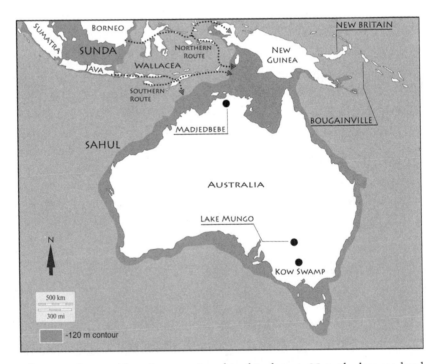

Figure 32 Sites and locations mentioned in this chapter. Note the low sea-level contour at −120m.

earliest modern humans as they arrived and dispersed across Wallacea and the myriad of islands of the wider region, and consider whether or not our ancestors could have met Denisovans in this key region.

First, I want to explore what we can deduce from the genetic evidence, both from Denisova 3 and from modern human genomes. David Reich invokes a nice analogy to explain how modern population genetics can shed light on the dispersal of modern humans and their interaction and admixture with other populations. Because the archaic genetic signal is divergent from modern human DNA it can, even in small proportions, help us to identify traces of gene flow, rather like medical imaging dye helps us to trace the movement of chemicals along human blood vessels.

In 2011 Reich's team at Harvard obtained genetic data from an additional thirty-three populations in Asia and Oceania for analysis, including Aboriginal Australians, Polynesians, Fijians, people from Indonesia and East Asia, and three so-called 'Negrito' groups: the Mamanwa from the Philippines, the Jehai of Malaysia and the Onge in the Andaman Islands.[3] The 'Negritos' are characteristically small in height, with a gracile build and dark pigmentation. Their physical appearance has been the basis for a suggestion that they may represent a relict population from an early out-of-Africa diaspora. Could they be amongst the descendants of the early modern humans who beat a path eastward towards Australia?

Reich compared 353,143 SNPs, those variable sections of the human DNA sequence, from areas of the genomes where there was high coverage between the modern people and the Denisova genome. The first observation made was that the Papuan and Australian groups have almost exactly the same proportion of Denisovan DNA. This suggests that Denisovan gene flow into these populations probably occurred *before* they dispersed into Sahul – the now partially submerged but previously joined continent of Australia and New Guinea. It also supports the idea that the two largely descend from a common ancestral population.

Denisovan ancestry is strongly correlated with Near Oceanian ancestry.* Fijians, for example, have 56 per cent of the Denisovan DNA of a New Guinean while having 58 per cent Near Oceanian ancestry. People from the island of Niue have 27 per cent of the Denisovan DNA proportion and 30 per cent Near Oceanian ancestry.

There were two exceptions to this, the Mamanwa and the Manobo from the Philippines. Reich tested several different models to try to explain why this was. The most likely reason was that after Denisovan gene flow into the ancestors of Australians, New Guineans and the Mamanwa there was a later admixture between the Mamanwa and east Eurasian modern humans, thereby reducing their Denisovan genetic component.

This suggested that there were two population movements of modern humans into Southeast Asia. First, a population of people who were the ancestors of modern New Guineans, Australians and Mamanwa, who interbred with Denisovans. Second, the ancestors of modern Indonesians and East Asians. This group did not appear to have admixed with Denisovans and in all likelihood derive from a much more recent dispersal event.

The populations that have Denisovan admixture today are in the east, and concentrated in Island Southeast Asia, and this suggests to me that it is more likely that the interbreeding event or events may have occurred there.[4] If true, this would imply that the Denisovans were spread across a wide range of different ecological and environmental zones in prehistory, from the colder, temperate regions of Siberia to the tropical islands of Asia and, as we saw in Chapter 7 with the Xiahe specimen, even onto the Tibetan plateau. If Denisovans had spread into these distant areas then they must have been able to adapt to very different environments and had a considerable technological and adaptive ability to survive. Modern humans have been thought previously to be the only humans capable of doing this (we encountered the term 'generalist specialist' to describe modern humans in Chapter 2). Perhaps once again we are significantly underestimating the abilities of our archaic cousins.

---

* Near Oceania is a term that describes the area that was settled by humans from around 35,000 to 40,000 years ago and includes the western islands of Melanesia, the Bismarck archipelago and the Solomon Islands.

It was later found that Denisovan ancestry was even more widespread across eastern Asia than previously thought, but at low levels.[5] Trace Denisovan ancestry was found in the Americas (<0.2 per cent), in Native American groups in both North and South America and amongst east Eurasian populations.

There seemed to be two likely explanations for this at the time. First, interbreeding occurred in a group of modern humans ancestral to all of these populations and then, later, those in the Americas and East Asia interbred with other modern humans, resulting in a lower or diluted proportion of Denisovan DNA. This might mean that interbreeding with Denisovans happened anywhere in eastern Eurasia. Second, interbreeding with Denisovans occurred in the population ancestral to New Guineans, Australians and Melanesians, and the trace Denisovan DNA in eastern Eurasians was derived via a population movement afterwards.[6]

Ancient DNA work could resolve this issue, because if high levels of Denisovan DNA were to be found in fossil bones in East Asia, this would support the first possibility. An opportunity to test this has come with the DNA analysis of an ancient human bone from a site in China dated to around 40,000 years ago. The bone comes from a collection of skeletal remains excavated at a site called Tianyuandong ('dong' means cave in the Chinese language, so it is translated as Tianyuan Cave), near Beijing and 6km south-west of the famous Zhoukoudian complex of sites.

Tianyuan was discovered in 2001 by local farm workers and later excavated by a team from the Institute of Vertebrate Paleontology and Paleoanthropology in Beijing. They found thirty-four skeletal remains from what was interpreted as an early modern human. Radiocarbon dating indicated an age of 39–42,000 years old.* The physical anthropologists who analysed the bones noted certain rather archaic features on some of the skeletal elements amongst the expected

---

* I think that the sample could be a little older. I tried to take a new sample of the bone once when I was in Beijing, intending to date it more reliably, but the curators in charge at the time did not give me permission, so I left empty-handed. My feeling is that the age we have currently is a so-called 'minimum age', and that it's likely that Tianyuan is *at least* 42,000 years old.

modern human-derived features. Could this be evidence for some kind of interbreeding? Only DNA could truly confirm this hunch.

The foot bones of Tianyuan were also interesting. The researchers compared the robustness of foot bones, particularly the middle toe bones, from Middle Palaeolithic humans dating to older than 40,000 years ago with those of modern humans from 30,000 to 40,000 years and with recent humans.* Bone is plastic and it responds to stresses and strains. The idea is that after humans began to wear shoes, the sizes of some of our foot bones decreased. People who are habitually barefoot have gaps between their big toe and the other toes, but when shoes are worn the toe bones tend to become more gracile and smaller while the leg muscles remain largely unchanged. The analysis showed that the middle toe bone of Tianyuan was gracile or delicate and they therefore suggested that it seems highly likely that this person wore shoes.[7] Tianyuan is the earliest evidence we have for this type of behaviour, which we think becomes more and more common from 25,000 to 40,000 years ago. Actual evidence for shoes, however, is exceedingly rare in the archaeological record.†

Researchers at the Max Planck Institute in Leipzig attempted to extract nuclear DNA from Tianyuan, but encountered a problem in their initial efforts. The difficulty was that the DNA was very poorly preserved: only 0.03 per cent of the DNA was endogenous or original to the bone. They used an ingenious new approach to get around this called 'bait capture' to enrich or increase the recoverable DNA and extract a sufficient amount for analysis. This method enables scientists to build large arrays of DNA sequences that they expect will match the sequences of the human DNA they hope to extract from ancient samples. It is possible, through commercial

---

* The modern people included a habitually shod Inuit man, an American from New Mexico (a cadaver) and a habitually barefoot prehistoric Native American man.

† The earliest direct evidence is a leather shoe that my team dated in 2009 from a site in Armenia called Areni Cave. The shoe dated to around 3600 BC. An earlier form of footwear, in the form of a sagebrush sandal, was found in the Fort Rock Cave site in Oregon, USA, which dated between 9,000 and 13,000 years ago. Only exceptional preservation conditions allowed these finds to be made.

companies, to design these probes to target the correct DNA sequences, but it is expensive. The team focused on capturing DNA matching the sequence of human chromosome 21 by using the sequences published from the human reference genome.* They used 973,304 probes to attempt to fish DNA from the Tianyuan samples and were able to map around 30 million bases of the non-repetitive section of the genome – a truly incredible scientific feat.

The results revealed that Tianyuan had similar amounts of Neanderthal DNA compared with modern people living in mainland Asia (around 1.7 per cent), but, at the resolution of the methodology applied, it was not possible to determine how much Denisovan DNA there was. As we shall see shortly, however, subsequent approaches were able to sharpen the resolution.

In March 2017 fascinating new evidence emerged from a team at the University of Washington in Seattle that shed new light on these questions of Denisovans and aspects of ancient admixture events in eastern Eurasia, Island Southeast Asia and Melanesia. They decided to probe the archaic parts of modern human DNA sequences using a powerful new statistical method known as the S* statistic.[8] This enables the analysis to be done in a much more rapid manner than was previously possible.† They explored non-African genomic data from the large 1000 Genomes Project database,‡ as well as from 5,600 living people from nineteen populations across Asia, Europe and the Americas.

When they looked at the results, they observed two clearly different components of Denisovan DNA in some of the modern populations they had studied. The first component, the so-called 'high-affinity' part, consisted of DNA that was related most closely to the Denisova 3 genome. They found this DNA in the genomes of living people in Japan and three populations in China (Beijing, Dai

* Chromosome 21 is the smallest chromosome and consists of 47 million base pairs (the largest, chromosome 1, is more than five times bigger).
† The analysis of archaic sections of DNA from more than 4,000 individuals in the UK 10K Biobase took less than four hours.
‡ www.internationalgenome.org/.

and Southern Han). The second, so-called 'moderate-affinity' part, was more distantly related to Denisova 3. This DNA was found in living Papuans, Oceanians and Melanesians. The Chinese and Japanese groups, however, had evidence for *both* the high-affinity and the moderate-affinity types. This confirmed the work mentioned earlier that had shown there was a tiny proportion of Denisovan DNA likely in East Asian and Native American populations, but at around 0.2 per cent it was twenty-five times smaller than the proportion of such DNA in Melanesians.[9]

This confirmed that there were once two different, genetically divergent groups of Denisovans. Rather than simply one homogeneous population spread across eastern Eurasia, there must have been a more complex structure in the Denisovan population. In contrast, the team found that the Neanderthal DNA in modern human groups from Europe and Asia appeared to derive from a *single* population and lacked the same underlying complexity.

There was another fascinating finding. Sections of archaic DNA were identified that could not be aligned to any archaic genome thus far found. This comprised around a *quarter* of the archaic DNA signal. It did not align with either the Denisovan or the Neanderthal genomes. The most likely explanation for this in my view is that it is Neanderthal and Denisovan DNA that can't be aligned simply because we have yet to capture the wide range of genetic variation in these groups due to the limited number of high-coverage genomes. A second and more intriguing possibility is that this unknown DNA comes from another archaic hominin, as yet unidentified genetically, that also introgressed into our ancestral human genomes thousands of years ago. We will come back to this in Chapter 14.

The discovery that a second population of Denisovans, more distantly related to Denisova 3, has contributed DNA to people in New Guinea and Melanesia invites us to look much more closely at this key emerging region in palaeoanthropology.

New Guinea is one of the most ecologically diverse islands on Earth. At 900,000km² it is the second largest island in the world after Greenland.[10] Its topography is complex, with a mountainous spine carving much of the island in two, peaks topping 4,500–5,000m in the central highlands, and wide floodplains in the lowlands and some of the central valleys. The climate is variable but dominated for the most part by tropical high-pressure systems, the presence of the inter-tropical convergence zone, that equatorial region of converging trade winds that is sometimes called 'The Doldrums', and an oceanic influence. Rainfall varies from 7,000 to 8,000mm per annum in the central mountains to 1,000mm in the capital of Papua New Guinea, Port Moresby. Its location on the Pacific Ring of Fire means that volcanic and seismic activity is common, and this has undoubtedly shaped land use and human geography from the time that the first humans arrived.

The vegetation and topography have doubtless also contributed to the amazing diversity in human groups and languages on the island. New Guinea is estimated to have more than 850 languages, constituting a remarkable 12 per cent of the total number of languages known in the world.[11] Some languages are spoken by thousands, some by mere dozens, but they have persisted because of isolation shaped by geography.

Is this diversity in tribes and language mirrored in the genetic diversity of people living on the island today, or are living New Guineans closely related to one another? Genetic studies here are in their infancy, but new results suggest that there are strong genetic distinctions between different human groups. A recent study examined samples from 381 living people from eighty-five different language groups, mainly in the west of Papua New Guinea.[12] The results showed that the differences between some tribal groups were much greater than those that exist between all of the people living in modern Europe. They were particularly pronounced between people living in the highlands and mountainous parts of Papua New Guinea and those in the lowlands. Genetic separation between some groups is estimated to extend as far back as 10–20,000 years ago. There were even significant differences between geographically close groups, no doubt maintained by the twin barriers of language and landscape.

Unfortunately, testing whether this genetic complexity extends back into deeper prehistoric time is going to be tricky, because of the old problem of DNA preservation in the tropics. For this reason, researchers have focused their efforts on expanding the analysis of DNA from living people in this region, rather like the team from the University of Washington did.

A team from Massey University in New Zealand has recently obtained DNA from 161 people from fourteen different island groups, spanning Sumatra to New Britain.[13] This included the first whole genomes ever sequenced from Indonesia, which is a sobering thought when one considers that Indonesia is the world's fourth largest country with 267 million people and a landmass the size of continental Europe.

The team were interested in exploring more about the two Denisovan populations that were previously identified. It turned out, however, that they found much more than they were bargaining for. They filtered out only the Denisovan parts of the archaic DNA in modern New Guineans (using powerful statistical methods including the S* approach) and compared these against the Altai Denisovan genome. By undertaking a 'mismatch analysis', they measured the number of times the Denisovan DNA in Papuans did *not* match with the Altai sequence. Unexpectedly, they found not one but two peaks of mismatched DNA. This could only mean that Papuans have inherited DNA from *two* genetically different populations of Denisovans.

They called these D1 and D2.

Now, recall that in East Asia, the Washington University team found that some people also have two introgressions, one of those is the same D2 as in the Papuans, while the other, D0, is different (D0 is the closest match to the Altai Denisovan). It seems, then, that the D1 population is restricted to Papuans alone.

This evidence therefore suggests that there may in fact have been *three* genetically different populations of Denisovans.

Interestingly, this emerging scenario also fits well with a reanalysis undertaken recently of the 40,000-year-old Tianyuan human we met earlier. Although, initially, no Denisovan DNA could be detected in Tianyuan, improved statistical approaches now show that there actually is a very small proportion of Denisovan introgressed DNA in him, equivalent to less than 10 per cent of the amount of Neanderthal DNA he had.[14] This Denisovan DNA is from a different source population than that found in Papuans and Aboriginal Australians and is more closely related to the Altai Denisovan population. This nicely confirms the evidence from the modern human genome work showing that there are indeed different populations of Denisovans in regions of eastern Eurasia and Southeast Asia, and that these contributed variably to the genomes of modern humans, and on more than one occasion. The highly fragmented Denisovan DNA in Tianyuan Man is probably derived from an introgression that occurred substantially before he lived – by how much is difficult to say on current evidence. Finally, Tianyuan appears to have come from a population that was related to most present-day Asians and also Native Americans, so thereby postdates the divergence between Europeans and Asians.[15]

It looks as though D1 split from the Altai D0 population around 300,000 years ago.* The D2 group seems to have split even earlier than this, in fact closer to the date estimated for the separation of Neanderthals and Denisovans. This implies that D2 is probably a third sister group to both. It appears to be as genetically different from other Denisovan groups as they are from Neanderthals.

Again, it's probably worth pausing at this point because we have certainly covered some pretty astonishing ground in the last page . . . not one but *three* possible Denisovan populations, the evidence mined from the DNA of *living* people. The possibility really is quite incredible. I have to scratch my head in wonder: after all it's only since 2010 that we even knew Denisovans existed.

Clues about *when* these various introgressions from Denisovans into us occurred can be mined from the genomic evidence. As one gets closer back in time to the introgression dates so the signal of the introgressed DNA is stronger and the size of those DNA blocks is bigger. Calculations based on block sizes suggest D2 introgressed into the ancestors of New Guineans around 46,000 years ago.[16] This

---

* The calculation ranged between 9,750 and 12,500 generations ago (280–360,000 years ago).

is in line with previous estimates which suggest that Denisovan admixture predates the population split between Papuans and Aboriginal Australians.[17] The D1 introgression into Papuans, on the other hand, is estimated to have carried on for longer – perhaps until around 30,000 years ago. This is a big surprise because, if it is a reliable estimate, it suggests that a very late admixture occurred in the regions around New Guinea. This might have occurred after the different modern human populations in New Guinea had become separated from one another.

By comparing the DNA of people living in New Britain, an island to the east of Papua New Guinea, and Papuans, the Massey University team showed that this is indeed probably what happened: there were differences in the relative proportions of the D1 population in at least two modern groups.

New Britain is thought to have been settled by *Homo sapiens* at least 35,000 years ago. It required an ocean crossing to reach there. There was less D1 DNA in the Baining people living on New Britain compared with Papua New Guineans, implying that there was an additional introgression of D1 into Papuans after they had become geographically separated. The split time between mainland Papuans and the Baining is estimated to be 16,000 years ago, which might well mean that the D1 Denisovan people survived until much more recently than once thought. The Philippine Negritos we met at the beginning of this chapter also record a second Denisovan introgression following their island separation.[18] This D1 population almost certainly lived east of Wallace's Line in the New Guinean region, I think, because none of their DNA is found in East Asian people today.

The earliest humans dispersing from Africa eventually made it all the way to Australia. To do this our human ancestors must have journeyed through or close to Wallacea and New Guinea. In fact, as we shall soon see, the New Guinean route is the most likely one. As previously noted, the earliest date for Australian settlement acts as a minimum age for an African exit. These early humans must surely have met Denisovans, based on the genetics and the introgression date estimates. What more can we find out about what happened when modern humans met Denisovans in Sahul? Is there any evidence for Denisovans aside from the genetic signals in living people? What does the archaeology of this period tell us about what happened and how people finally reached the last station on the out-of-Africa line: Australia?

Like New Guinea, Australia is linguistically and culturally complex. It is thought that, at the time of initial European contact, there were approximately 250 spoken languages, from twenty-eight language families. Of these families, twenty-seven are restricted to the far north, while the remainder are dominated by one family, called Pama–Nyungan. These cover 90 per cent of the country and constitute the largest hunter-gatherer language group in the world. It seems likely that this dominating family dates from, and spread within, the mid-Holocene period, 6–5,000 years ago.[19] The distribution of other language groups prior to this is not well known.

The earliest evidence for human beings in Australia fits broadly between 45,000 and 65,000 years ago. The wide range reflects the arguments of scholars who favour a series of early dates from a small number of archaeological sites obtained using optical methods of dating,[20] and those who do not.[21]

People moving into Australia from Sunda had two options: a northern route through New Guinea or a southern route via Bali and Timor to north-western Australia. Modelling work has indicated that the northern route is the most likely.[22]

Our understanding of the earliest settlement of Australia has been hugely influenced by a series of archaeological discoveries of human remains in the twentieth century. One of the most important occurred in August 1967 when Alan Thorne (the Australian physical anthropologist we met in the previous chapter expressing doubts about the Hobbit) was examining human skeletal remains in the National Museum of Victoria in Melbourne. He opened a drawer and came across some bones of what looked like a remarkably robust human being.[23] There was a distinctive carbonate-like crust on some of the bones that reminded him of the appearance of a famous cranium from a site called Cohuna, which had been found in 1925 by a farmer ploughing a field at the edge of a large Victorian wetland called Kow Swamp. The cranium was also quite robust and distinct.

Thorne looked into the museum records associated with the skeletal remains and found that there was a police report relating to it.* Through the police records he was able to deduce that the bones he had spotted came from very near the site where the Cohuna skull had been found.

The following year Thorne began excavations at the site. Remarkably, he found the remaining parts of the same partial skeleton, including the other half of the bone he had found in the museum drawer.[24] Subsequently, more archaeological investigations yielded other human remains from around the margins of the swamp, and by 1972 the remains of around forty humans had been identified.

Some of the remains were buried with offerings of mussel shells, the teeth of marsupials and sprinklings of red ochre or haematite (iron oxide). Dates indicated that people were living at Kow Swamp 19–22,000 years ago, when temperatures globally fell to their lowest levels in the last 100,000 years.[25]

Thorne interpreted these human remains, and others like them, as having the 'mark of ancient Java' due to their robust cranial shape. What he meant was that he thought there were great similarities between the Kow Swamp individuals and the remains of *Homo erectus* from the Indonesian island of Java excavated earlier in the twentieth century. Thorne thought that modern humans had evolved separately in different parts of the world – the so-called multiregional model of human origins. This model was in direct opposition to the Out of Africa model. In Australia, he thought, humans had ultimately evolved from East Asian *Homo erectus*. Later this was modified to include a model that involved greater gene flow to explain the close similarities between different modern humans in different regions of the world.

In 1969 other excavations produced more key evidence for early

---

* Often, when human remains are found it is usually the police who are involved first, then archaeologists if it has been established that the skeleton is likely to be old and not forensically important.

Australians. At the Willandra Lakes, a semi-arid lake-and-dune system in New South Wales, human remains with a more gracile appearance were discovered. The first human skeleton to be found by archaeologist Jim Bowler was WLH1 (Willandra Lakes Hominin 1), aka Mungo Lady, a young female.[26] Later, in 1974, the skeleton of a man lying on his side was found around 400m away. This burial (WLH3) had evidence for being interred with sprinklings of red ochre – amongst the earliest evidence ever found for this type of burial behaviour. Dating of the sediment in and around the graves subsequently yielded dates of 41,000 ± 4,000 years ago.[27]

Thorne interpreted this as showing that Australia was settled by two different populations: an initial gracile, perhaps modern human population, and a later more robust group, more like *Homo erectus*.[28] The idea was that modern Aboriginals originated from the hybridization of these two groups. He termed this the di-hybrid hypothesis.

Others were not convinced. They thought that both robust and gracile groups were actually within the range of modern Aboriginal people and wider present-day humans, and that there was no strong biological link to Indonesian *Homo erectus* at all.[29] Some of the Kow Swamp crania were argued to have been affected by cranial deformation during childhood, which had exacerbated their rounder shape.[30] This is the view now amongst most specialists.

Given what we now know about multiple Denisovan populations and interbreeding, though, I wonder whether some of the human remains from Australia might in fact reflect evidence for ancient admixture on that continent. Could DNA help to address this?

Extracting endogenous DNA from human remains in the warmth of Australia is very challenging. Previous genetic work resulted in mtDNA being extracted from some of the key human skeletal remains, including WLH3 and KS8 from Kow Swamp,[31] but these sequences were shown later to be severely contaminated and therefore unreliable.[32]

In 1990 the Kow Swamp human remains were quite rightly returned to the local Echuca Aboriginal community.[33] They were reburied in a mass grave whose location is known only to the local people. Mungo Lady returned to her people in 1992.[34] Later, in 2017,

Mungo Man came home as well. One of the tribal elders said that 'after 40,000 years this man came up from the ground to walk with us for a while. And he's told the world that we are a very ancient people.'[35]

The genetic and archaeological evidence described in this chapter shines a new light on how early modern humans may have spread into the vastness of the southern continent of Sahul. I think it is increasingly likely that once humans had crossed Wallace's Line and entered Sahul, they would have encountered Denisovans. The distribution of Denisovan DNA in modern people is concentrated east of Wallace's Line and the evidence we explored earlier in this chapter provides additional support for there being Denisovans probably already living in the New Guinea region and surrounding islands. These people may have been already well adapted to local tropical conditions by virtue of their longer stay in the region. One of the advantages of long-term life in the tropics is the benefit of greater immunity to some of the endemic tropical diseases there, as we will see in Chapter 16.

We know that the introgression of Denisovan DNA had occurred by around 46,000 years ago, before Australian Aboriginals and Papuans split from one another.[36] We can also tell from the average lengths of Neanderthal and Denisovan DNA in Australo-Papuans that the time that has elapsed since Neanderthal admixture is longer than the equivalent time for Denisovan admixture, by around 11 per cent. This supports the idea that the Denisovan genetic pulse or pulses introgressed into modern humans later than the Neanderthal pulse.* This isn't surprising, given the geographical distribution of both groups and the order in which early modern humans would have encountered them as they moved out from Africa and across Eurasia.

Movement into Australia from Sunda would have required watercraft; without a vessel of some kind it would have been impossible. It may be that modern humans adapted fairly rapidly to living in

* Estimated to be 55–84,000 years ago. See note 17.

maritime contexts once they arrived on the eastern margins of Island Southeast Asia. Evidence for very early fishing, before 40,000 years ago, has been discovered in archaeological sites on some of the islands of Wallacea.[37] Perhaps the people living in this region became accustomed to fishing, using boats and living a marine life, and these technological developments enabled them to expand on to Australia. This possibility is supported by demographic estimates and spatial modelling work, which suggest that the founding population of modern humans that made it to Australia numbered between 1,300 and 1,550 individuals.[38] This makes it more likely that groups arrived through multiple crossings: estimates suggest perhaps more than 130 people at a time over 8–900 years.[39]

The earliest evidence for human settlement in Australia has been found at Madjedbebe Rockshelter in northern Australia, from about 65,000 years ago,[40] perhaps even earlier, although this has been criticized.[41] The people who lived there used grinding stones to process a range of plant and seed foods, they had pigments, mineral crayons and advanced stone-tool technologies. This early date might suggest that the people at Madjedbebe did not contribute anything to the DNA of modern Aboriginal Australians, because it appears to predate Denisovan introgression. There are significant uncertainties in both the DNA estimates and the archaeological dates, however, which means they may well overlap.[42] I think it is more likely that, following initial modern human dispersal into Sahul, admixture occurred and then, shortly afterwards, humans moved into Australia. If this is true, then Madjedbebe would provide a minimum age for human interaction with Denisovans.

There is an alternative, however, and that is that modern humans were not the first people to disperse into Australia. As we have already seen, the Hobbits and *Homo luzonensis*, or their ancestors, also crossed Wallace's Line. If Denisovans were in New Guinea then they also managed it,[43] and if they were there, then they were as good as in Australia, since Sahul was one joined up super-continent, as we have seen. We cannot identify Denisovan presence based on tool technology yet, so we will have to rely on future archaeological work to shed further light on this. Given the evidence we have already about the extensive geographic distribution of Denisovans – from the Siberian Altai to the Tibetan plateau to the drowned margins of Taiwan and Island Southeast Asia, how could we not at least entertain the possibility that Denisovans also made it to Australia?

Of course, we will never know the motivations of these first human colonizers of Australia, whoever they were, and the reasons they chose to undertake these kinds of ocean crossing. It is easy to invoke the spirit of human adventure and exploration, but people do not usually chance their lives unnecessarily in ventures like this; they prefer a calculated risk. I think the first Australians made deliberate crossings, in number, based on a recent history of maritime adaptation. They founded a population that gave rise to one of the longest and most successful human adaptations outside Africa, a settlement that was finally, catastrophically interrupted by the arrival of the first Europeans in the seventeenth century.

14. *Homo erectus* and the Ghost Population

There are other mysterious hominins that have lurked in the background of our story thus far. We saw in the previous chapter that a proportion of the archaic DNA of living people in Island Southeast Asia and Melanesia derives from an unknown source, a hominin that has yet to be genetically sequenced. Another mystery archaic hominin contributes between 2.7 and 5.8 per cent of the genome of Denisova 3, a hominin that diverged from the others 0.9–1.4 million years ago.[1] Geneticists call these tracts of DNA the remains of 'ghost populations', because we do not yet know whom they belong to in the fossil record.

Who could these hominins be? I think one strong potential candidate is *Homo erectus*, one of the most intriguing and long-lasting members of our human family. In this chapter I want to explore this further and see where the evidence leads us.

There is a long history of hominin discoveries and archaeological research in East Asia. In the 1800s the influential German zoologist Ernst Haeckel was convinced that the origin of humans lay in this region. Unlike Darwin, Haeckel favoured the east owing to the presence of similar primates, such as the orangutan and the gibbon. In a

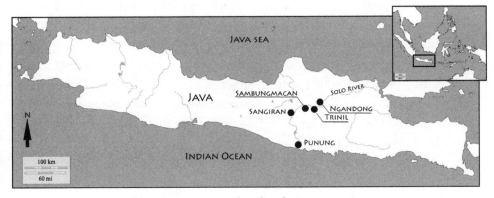

Figure 33 Sites and locations mentioned in this chapter.

rather confident move, he even coined a new name for the purported 'missing link' that he felt would one day be found: *Pithecanthropus alalus* (apeman without speech).

A young Dutch doctor named Eugène Dubois was so influenced by Haeckel, and so excited by the new scientific research into human origins then happening, that he decided he must somehow become involved. As a young man, Dubois had been fascinated by the natural world. He had heard about the Neanderthals because in July 1886 their fossil remains had been found near to where he lived at a site in Belgium called Spy.[2] He decided that he must somehow go east, join the search for human fossil remains and test Haeckel's suggestions that a 'missing link' lay there.

Then he came up with a brilliant idea: he would seek an army posting to the Dutch East Indies.

So it was that in 1888, at the age of twenty-nine, he arrived on the Indonesian island of Sumatra. In his spare time, he excavated several cave sites, but the sediments and archaeology in them were too recent in age. Despite this, his success in finding ancient bones in the region resulted in his being relieved of his army post and rewarded with a team of workers to help in his scientific research efforts. Seeking older sediments, he moved to the island of Java. He decided to dig on a small bank protruding into a bend in the River Solo in central Java.

He was in luck. His workers almost immediately began to find fossils of very old extinct animals. A month later, in September 1891, they found a tooth. It was the upper molar of an ape-like primate. Dubois was ecstatic, but identifying the specimen would be tricky. What he needed was a comparative collection of modern primate and human skulls in order to identify whether his new specimen was something entirely new. He sent letters to colleagues back in Europe requesting that a chimpanzee skull at least be sent to the East Indies, but it was a very slow process. He had no choice but to wait.

In October, however, he found something that eclipsed the tooth. It was the chocolate-coloured skullcap of a clearly big-brained primate. It came from the same general area in the site as the tooth, and Dubois thought they must derive from the same individual. He recognized quickly that it was different from a chimpanzee, as well as from the Neanderthal remains then discovered. But again, hamstrung by the lack of any comparative material, he could not give the new finding a scientific name.

The following year his team of workers found a beautifully preserved right femur, very similar to that of a modern human. Dubois said, 'I can't help thinking that there is a lot of ape in that skull, and a lot of man in that femur. Doesn't that sound like a transitional species, a link between apes and man?'[3]

By late 1892 he had the long-awaited chimpanzee skull, along with skulls of a gibbon and modern humans. He threw himself into anatomical analysis and comparative measurement. Finally, after weeks of work, he concluded that there was enough evidence that he had something entirely new on his hands: a new species of human. He decided that his specimen would be named *Pithecanthropus erectus* – *Pithecanthropus* in honour of Haeckel, and *erectus* due to its human-like femur. The apeman who walked upright.

Dubois had pulled off a scientific miracle.

Today we call this species *Homo erectus* to reflect its position in the genus *Homo*. In Island Southeast Asia, *Homo erectus* appears in the fossil record as early as 1.3–1.5 million years ago.[4] Their kind must have been tremendously successful, because we find their hand axes and skeletal remains, from perhaps 100,000 years ago in Java all the way back to around 1.8 million years ago in Africa.[5] Curiously, despite these probable African origins, the majority of the fossil remains from this species are from Eurasia, raising the possibility that this hominin might have an origin outside Africa.[6] As always, dating is crucial, but challenging – at some sites, particularly challenging.

One of the key sites of *Homo erectus* is the Javanese site of Sangiran. The German-Dutch palaeontologist Ralph von Koenigswald discovered it in the mid-1930s. From a young age he had been inspired to work in the area from reading the story of Dubois.* Over the course

* Von Koenigswald's old teacher, Professor F. Broili, had had an enquiry from the Dutch Geological Survey about whether any of his students would be interested in going to Java to work as part of their palaeontological team. Von Koenigswald leapt at the opportunity and arrived in Java at the age of twenty-nine, the same age as Dubois had been.

of several years in the 1930s and early 1940s he made a series of stupendous fossil discoveries in the region, including several crania of *Homo erectus* that were similar to the *Pithecanthropus* specimen that had been found by Dubois himself. Work in Java came to an end when the Second World War reached the area, however, and von Koenigswald was captured by the Japanese. He spent thirty-two months as a prisoner of war, learning to read hieroglyphics as a way of avoiding boredom.[7] Fortunately for science, his wife Luitgarde (Lütti) and some of their friends rescued his precious fossil specimens and kept them safe until the end of the war.[8] Lütti kept his famous Sangiran IV upper jaw in her pocket throughout the years of Japanese occupation (it was, von Koenigswald said, his most important discovery).

More than a hundred hominin remains from the site have been found and they fall into two broad groups, one early and more primitive with parallels to early 1.4–1.7-million-year-old African *Homo erectus*[9] (which are often instead called *Homo ergaster*) and one later, after 780,000 years ago, which exhibit a larger cranial size. This generally applies to the remains of this hominin through time in Java – gradually their cranial capacity and skull shape changes, becoming larger in capacity and less robust in terms of facial musculature. The Javan skulls had similarities with others found in China.[10] Sangiran 17 is the most complete *Homo erectus* ever found in Asia. This was the specimen that was compared with the Kow Swamp skulls we met in Australia and argued to be morphologically very similar.[11] Sangiran remains a rich and important fossil locality to this day.*

We have a pretty good idea, then, that *Homo erectus* was outside Africa very early, but for me a more pressing question is when did *Homo erectus* disappear? It would be fascinating to find out, because if there is a possibility that they survived very late – into the last 100,000 years, for example – then they may have met or interacted with our early modern human ancestors. In addition, a late date such as this

---

* In 2016, eighty years after von Koenigswald's first discovery, a local villager called Setu Wiryorejo found another *Homo erectus* skull in rocky ground at Sangiran. He removed it with a crowbar before he realized what it was, and then donated it to the local museum. They paid him US$1,800 as a reward.

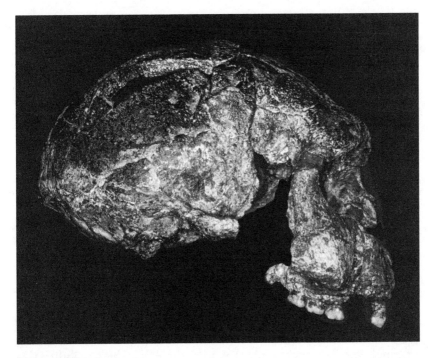

Figure 34 Sangiran 17.

would mean that they probably also overlapped with some of the Denisovan populations that may have been in Island Southeast Asia. Once more it is not hard to imagine what a very strange and curious experience this would have been for both groups, encountering a different species of people, separated evolutionarily from one's own for hundreds of thousands of years before being reunited on a distant island. I imagine these African exits of successive groups belonging to our genus as being rather akin to waves crashing on a beach. The farthest easterly margins of that first wave crested in this wonderfully complex location in Island Southeast Asia; later, could it be that a second wave of the human family washed up in the same location and there they met?

In 1996 it looked as though evidence had been obtained that supported such a late survival. Two wonderfully named American geochronologists, Carl C. Swisher III and Garniss H. Curtis, were working on dating the key site of Ngandong on the Solo, just 10km from Trinil, the site where Dubois had made his discovery of *Pithecanthropus* more than a century before. The site was excavated in the

1930s by the Dutch, who had found a plethora of *erectus* human remains, including twelve calvaria (skullcaps). Swisher and Curtis excavated a test pit of 1m² near the Dutch excavation area, in an attempt to find material to better date the deposits.[12] Down around two metres they found some fossil animal teeth. Assuming them to be from the same level as the *Homo erectus* remains, and therefore of the same age, they applied two different methods to date the teeth.* The results were unexpectedly young: between 27,000 and 53,300 years ago.[13] In their paper the authors said, 'The new ages raise the possibility that *H. erectus* overlapped in time with anatomically modern humans (*H. sapiens*) in Southeast Asia.' I remember the dates being published with a mixture of disbelief and excitement. Could it really be that *Homo erectus* was this young?

Some of my colleagues were not convinced. One of the statements in the original Swisher paper is revealing to read in hindsight. Of their small, 1m² excavation they say that 'the upper 2m appeared to represent colluvium, possibly including some of the debris from the original 1930s quarry'. In other words, there was the possibility that some of the material in the square was not in its original location, but was the backfill of the earlier Dutch excavators. If true, this would mean that the teeth, and therefore their dates, were probably not related to *Homo erectus* from Ngandong at all.

Confirmation of this came much later, in 2020.[14] An international team restarted excavations at Ngandong between 2008 and 2010 to establish a more reliable age for the remains of *Homo erectus* at the site. They discovered that because the Dutch had excavated a very large area of the total site (around 6,300m²), it seemed highly likely that the location of the Swisher and Curtis test pit was indeed located in their

---

* Electron spin resonance (ESR) and U-series dating were the methods used. ESR is a trapped charge method. Naturally radioactive particles in sediments (principally uranium, potassium and thorium) emit low-level radiation. Certain minerals in the ground react to this ionizing radiation and electrons within their crystal lattices move to higher energy states. Dating is possible when one measures this accumulated energy and compares it with the so-called 'dose rate', which is the background radiation present in the sediment. In archaeology, the method is primarily used on tooth enamel.

backfill. Their very late ages could therefore be discounted because of the possibility that the teeth they had dated were just very young teeth jumbled up in sediments previously excavated and then backfilled into the excavated holes by the Dutch when they had finished their work.

The team realized that to date the site accurately they would have to understand the way in which the sediments and the human calvaria within them had been deposited. This was not an easy task. The river plays a key role in both the deposition and the erosion of sediment in the region, from about half a million years ago. The Ngandong human remains themselves were deposited in a single flood event, from upriver, perhaps a result of changing environmental conditions. They are therefore what we would call secondary deposits, material that has been moved from its original location. This type of event makes dating difficult. How can one tell if the entire sequence of sediments and bones is not jumbled up as it is being redeposited?

The team carefully analysed the bones and teeth found from the same levels as the *Homo erectus* remains. They showed a distinct lack of damage. Some fragile bone material was identified and even soft tissues that appeared to have survived against the odds in some cases. This suggested that a single event was probably responsible for the deposition of material at Ngandong and that it was a low-energy event rather than a high-energy and very damaging one. (A high-energy event in this context would be something like a powerful flash flood.)

Luminescence dates obtained for the sediments suggested it was likely that the human remains and the animal bones were deposited at broadly the same time. The results together suggested an age for *Homo erectus* at Ngandong of 108–117,000 years ago,[15] not nearly as young as the discredited Swisher and Curtis results, but still surprisingly recent. These ages would imply that *Homo sapiens* and *Homo erectus* probably did not overlap in Southeast Asia, since the earliest evidence for our species there is significantly later (around 45–65,000 years ago). But, intriguingly, it would imply that Denisovans may well have met and interacted with *Homo erectus*.

The new dates allow us to work out how *Homo erectus* may have lived and survived at this time, and what environmental conditions were like.

From 90,000 to 120,000 years ago the sea levels around Java were up to 60m lower than today. The Sunda shelf was exposed and passage was possible on foot across a vast tract of now drowned land. Primates such as orangutans cannot cross water barriers, so their presence on Java demonstrates beyond doubt that islands like this were once joined to mainland Southeast Asia. Later, sea levels rose, the climate became warmer and wetter and the vegetation on Java changed in response. Instead of being dominated by open woodland, which is the type of environment that *Homo erectus* appears to have lived in, tropical rainforests spread. The ancient fauna gave way to animals more familiar to us: macaques, gibbons, tigers, pigs, deer, elephants and orangutans.

One of the most important sites that documents the changes in the animals of Java is called Punung. In the 1930s von Koenigswald worked here trying to build a picture of the rainforest environment and the animals that were present at the time of *Homo erectus*. Amongst the animal bone remains in Punung, he found something unexpected: five human teeth.[16] In 2001 researchers attempted to find these teeth in the von Koenigswald collections from the Punung site in Frankfurt. Sadly, they located only one, a rather gnawed premolar labelled '*Homo sp.*'

Analysis of the tooth has shown that it falls within the range of *Homo sapiens*. Although we must be cautious, it might just be, then, that our species was present on Java much earlier than hitherto thought.[17] As we have seen before in Niah Cave (Borneo) and in a site in Sumatra called Lida Ajer,[18] modern humans appear able to adapt to rainforest environments. Perhaps *Homo erectus* was not as well adapted. Perhaps, as the climate changed and sea levels rose, as the open woodland environment declined and rainforests spread, *Homo erectus*, that most successful of hominins, finally met its end in the far reaches of eastern Asia and our species moved into the available niche. Whether or not our species had anything to do with the disappearance of *Homo erectus*, at least on the basis of this evidence, is unclear. More work, and more fossil remains, are required before we can be confident. It seems odd to me, however, that, after such a long period of being present on Earth, *Homo erectus* would suddenly disappear just like that.

Ngandong is not the only late *Homo erectus* site in Indonesia. There is another, the site of Sambungmacan. Like Ngandong, and all of the other key *Homo erectus* sites, the sites attributed to Sambungmacan are once again on the Solo river in Java. (I say 'sites' because there are several places that have yielded human fossil remains in this locality.) The location was discovered in the 1970s and an almost complete skull was found, called Sambungmacan 1, or SM1. In 1977 another cranium was found (SM3).[19] This one was illegally placed on the antiquities market twenty years later, and found its way to a shop in New York City called Maxilla and Mandible, Ltd. specializing in fossil material. There it was identified as a missing Indonesian specimen and later returned to the country.[20] Like at Ngandong, dating the Sambungmacan material has been very controversial, with a range of methods applied resulting in a very wide range in age estimates. Recent work I have been involved in, however, has suggested the possibility of a young age for Sambungmacan 1, and therefore for *Homo erectus*.*

The Sambungmacan 1 specimen is one of the only *Homo erectus* crania that we have a good idea about in terms of its exact find spot. It was discovered by workers digging a new canal along the Solo in 1973. Returning to the site in 2005, members of our team discovered the find location and observed that below it were two layers of organic material, the lower of which contained around 2cm of peat-like organic material. This was radiocarbon dated to approximately 49,000 years ago. Previous work on dating the skull directly, using a technique called gamma-ray spectrometry, had produced a series of age estimates that overlapped with this. New analysis of the results suggests that we can have a good degree of confidence in these dates.

* The research group is led by Chris Turney of the University of New South Wales.

Statistically combining these various data gave us a result of between 50,000 and 53,000 years ago for Sambungmacan 1.[21] If reliable, this would put late *Homo erectus* clearly within the range of early *Homo sapiens* in Java and raise the possibility of overlap and interbreeding.

To really clinch this, however, we need something more: evidence for actual interaction in the form of genetic evidence showing interbreeding, or evidence for cultural exchange of some kind. As we have seen again and again, DNA analysis of ancient bones in the tropics is virtually impossible, but one possible avenue is to explore the DNA of modern humans for evidence of super-archaic DNA that could come from *Homo erectus*.

Our team therefore scanned genetic data from more than 400 populations worldwide, with around half of those from people living today in Island Southeast Asia and New Guinea. Denisovan and Neanderthal DNA tracts were removed, leaving between fifteen and eighteen MB (megabases) of DNA that could derive from other archaic sources (compared with around 12.5MB in other Eurasian populations). To give you an idea of the size of these chunks of DNA, when we look at the total DNA from Denisovan and Neanderthals in a person from New Guinea, we are looking at around 239MB,[22] so this is a very small additional proportion (*c.*0.6 per cent of the total). It looks promising, but in fact we think that the higher proportion in Island Southeast Asian and Papuan populations is likely to be attributed to DNA that actually derives from Neanderthals or Denisovans but that is undetected by the statistical methods being applied. Once this is taken into account, we found no difference in the inferred proportion of super-archaic DNA in Papuans or Australians compared with any other population. The results at present would therefore suggest that there is no evidence for *Homo erectus* interbreeding with early *Homo sapiens*. If *Homo erectus* were present until 50,000 years ago in Island Southeast Asia – and, as we have seen, the evidence is still emerging – then they do not appear to have interbred with us.

One final word of caution is needed about the term 'ghost population'. We ought to remind ourselves that the concept of a ghost population arises from blocks of DNA that are not able to be aligned against existing sequences, and thereby taken to reflect the likely presence of a population yet to be identified genetically in the fossil record. These ghost populations are identified using statistical programs that analyse admixture clusters,[23] so they are the product of statistical tools and therefore come with a degree of uncertainty. Ghost groups may well exist, but they need to be verified in the field with ancient bones and DNA analysis before we can be completely sure that they are real and not statistical constructs. In addition, we need to recall again that ghosts may simply be a reflection of missing data in the genomic variation in our own species; they may disappear as we map more and more modern and ancient genomes.

We are getting tantalizingly closer to deciphering what happened to the waves of human ancestors who left Africa and broke onto the beaches in the farthest parts of eastern Eurasia. It does not look as though there is any super-archaic ancestry in living humans that one can link with *Homo erectus*. If correct, this implies that we might have just missed each other in Island Southeast Asia. Denisovans, on the other hand, appear to have genetic evidence for DNA from a group that has not yet been sequenced (a ghost population). Could the introgressing population be *Homo erectus*? The dates from Ngandong suggest that this most successful human ancestor lived at least until around 100,000 years ago and therefore was highly likely to have overlapped with Denisovans. As we find out more about these different Denisovan populations and their geographic distribution, we may well find the demographic clues to when this might have happened. Given what we have already seen so far in this book, with diverse human groups meeting and interbreeding, I think that if the dates are correct it becomes highly likely that there was some interaction and gene flow between them. Time will tell.

15.   Disappearing from the World

It is a sobering thought, when we consider the tremendous range and variety in the human condition, that now, it's just us. As Carl Sagan said, 'Extinction is the rule, survival is the exception.'[1] We see so many extinct forms of life in the fossil record that it is clearly the common outcome for the majority of species that have existed on our planet. None of the human relatives we have met in this book that appear to have co-existed with us until 40–50,000 years ago are now present, from two or three different populations of Denisovans, to Neanderthals, Hobbits, *Homo luzonensis* and possibly also *Homo erectus*. We are now alone in the world, the only remaining twig of a genus called '*Homo*' that first appeared in Africa sometime after 2.5 million years ago.

How and why did this happen? Was the disappearance of the other groups to do with some genetic, cultural or technological advantage that we have that they did not? Did disease or a natural disaster play a part? Or was it all simply down to luck?

Of all of the extinct members of the human family we have met so far, Neanderthals are the best known, and we know more about their disappearance than about any of the others'. In the case of Denisovans, they are only recently discovered and we are still very much feeling our way in terms of their geographic distribution, dates and cultural adaptation. The lack of human remains means that at present any idea of a disappearance date is necessarily coarse. Yet there is still something we can say, based on the genetics that have been undertaken in both ancient and modern cases, as we will shortly see.

In the case of the Hobbits, we also know very little. The question of overlap or contact between the two populations has yet to be confirmed. My view is that it seems more likely that Hobbits disappeared upon or after the arrival of modern humans. Recent work tends to

1. Looking down on the Anui river valley and at the buildings of the Denisova Cave base camp, Siberian Altai. The cave itself is to the far right, hidden on the valley floor beneath the hills.

2. The entrance to Denisova Cave. Inside the cavern one feels a profound sense of awe thinking about the discovery here of a hitherto unknown kind of human, the Denisovans, a group that contributes to modern humanity genetically and perhaps culturally too. Tens of thousands of years have passed since their kind disappeared from here and from the world, but one still feels their presence in the shadows.

3. The author (*centre*) in the East Chamber at Denisova Cave working with Michael Shunkov and Maxim Kozlikin (*front*) and Katerina Douka to obtain samples for dating the site.

4. Neanderthals appear to have had a genuine interest in feathers. They targeted the wings of large raptors and perhaps decorated themselves with them. At the site of Grotta di Fumane, Italy, and in Neanderthal layers around 44,000 years old, archaeologists found many cut wing bones from birds of prey such as falcon and lammergeier, as well as Alpine chough and wood pigeon. This reconstruction is based on that work.

5. At Gorham's Cave, Gibraltar this enigmatic 'hashtag' was found carved into the rock. Archaeological evidence supports a Neanderthal hand in making it.

6. Is this Neanderthal art? This red scalariform image from La Pasiega in Spain dates to more than 64,800 years ago; a time when we know of no modern humans present in this part of Europe.

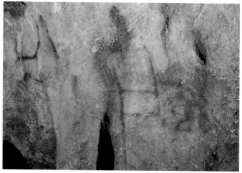

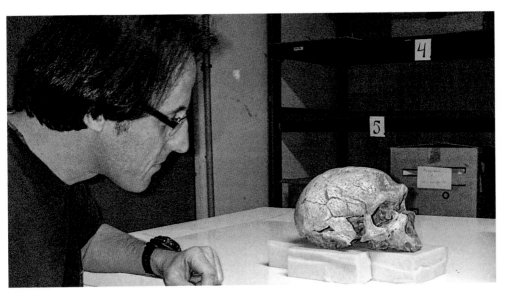

7. The author in the basement of the Musée de l'Homme in Paris meeting the 'Old Man' of La Chappelle-aux-Saints. The reconstruction of this skeleton in the 1910s led to the widespread impression that Neanderthals were brutish, squat and primitive. In fact, he was 40–45 years old with poor dental hygiene and some degenerative joint disease. He was probably cared for by other kin or family, just as we would look after one of our older family members today.

8. Johannes Krause was working in his lab in Leipzig on the DNA sequence of a tiny piece of bone from the Russian Altai when he noticed something unusual. It turned out that the DNA came from a completely new species of human; the Denisovans. Not surprisingly Krause remembers this as 'scientifically the most exciting day of my life'.

9. Svante Pääbo, of the Max Planck Institute for Evolutionary Anthropology in Leipzig, Germany. As a pioneer in ancient DNA Svante led the Neanderthal Genome Project and his group discovered the Denisovans in 2010. Svante is holding the Ust'-Ishim femur, a bone that was found in Siberia by mammoth ivory hunters. Svante's team found large chunks of Neanderthal DNA in Ust'-Ishim's genome which showed that his ancestors had interbred with Neanderthals around 55,000 years ago.

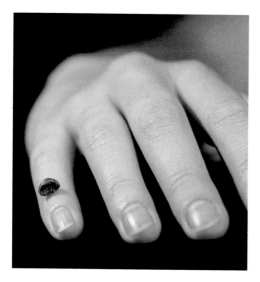

10. Denisova 3. The tiny pinky-finger bone whose DNA sequences stunned the world in 2010.

11. Artist's reconstruction of what the Denisova 3 girl may have looked like.

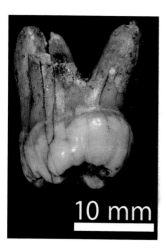

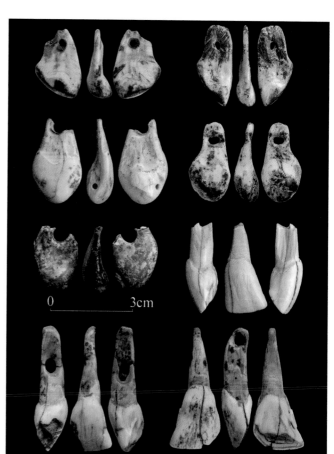

12. Denisova 4, the massive Denisovan tooth found in 2000 in the South Chamber of Denisova Cave.

13. Pierced mammal tooth pendants from Denisova Cave.

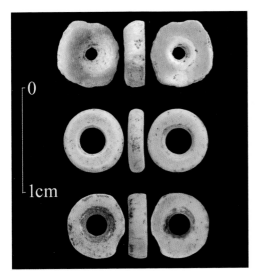

14. Ostrich eggshell beads excavated at Denisova Cave.

15. A Palaeolithic flute from the site of Höhle Fels, Germany dating to ~40,000 years old. It was found in the summer of 2008 in twelve pieces in the Aurignacian level of the site, so it was made by *Homo sapiens*. The instrument was made from the radius or wing bone of a griffon vulture; a bird which has a wing span of over two metres.

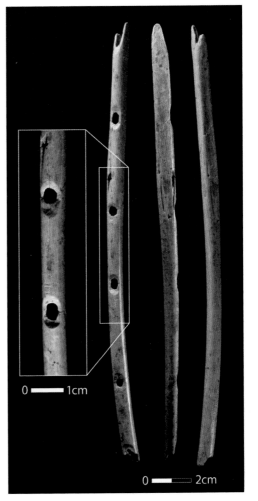

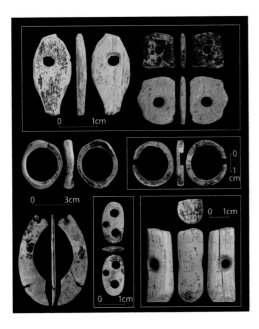

16. Rings, beads, plaques and a possible button (*bottom right*) made out of mammoth tusk from Denisova Cave's Initial and Upper Palaeolithic levels. Were these and the other ornaments here made by both Denisovans and modern humans?

17. 'Denny' was found using ZooMS, a method of detecting human bone based on collagen fingerprinting. The tiny bone, just under 2.5 cm long, derived from a first-generation offspring of a Denisovan dad and a Neanderthal mum (this is a 3-D print of the original bone).

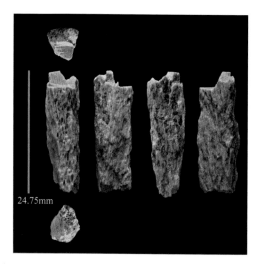

24.75mm

18. Smooth surfaces on the bone indicate that the bone probably passed through the gut of a hyaena.

19. Sam Brown with a bag of bone fragments from Denisova for ZooMS analysis. On average, one in a thousand bones in a bag like this turns out to be a human bone. Denny was one of them.

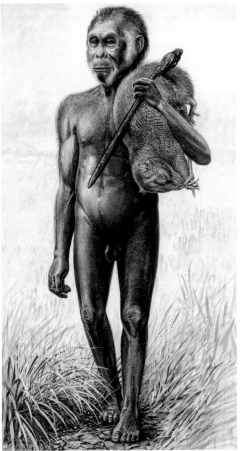

20. In 2003 at the cave of Liang Bua (*top*) on Flores. Thomas Sutikna (*above*) and Benyamin Tarus work to excavate the skeleton of LB1, the Hobbit (reconstructed on the right by artist Peter Schouten). They initially thought the skeleton was that of a child but soon afterwards an analysis of the teeth showed they were all erupted and some were worn, a clear sign that LB1 was an adult, albeit an adult of Hobbit-like proportions.

21. The Xiahe jawbone. In 1980 it was taken from Baishiya Cave by a monk after prayer and given to the leader of his monastery for safekeeping. Almost thirty years later scientists showed it was a Denisovan using cutting-edge proteomic analysis. It is the first Denisovan human bone to be discovered outside the eponymous site (note the absence of a chin, a key sign it is not *Homo sapiens*).

22. Baishiya Cave on the Tibetan Plateau in western China, the holy cave that furnished the Xiahe Denisovan jawbone shown above. Excavations continue at this high-altitude site today. This shows that Denisovans had been at the site more than 100,000 years ago and perhaps as recently as 45,000 years ago.

23. In 1990, after digging on and off for three years, a teenager called Bruno Kowalszcweski discovered a cave in the Aveyron region of France that contained something extraordinary. More than 170,000 years ago, Neanderthals had been deep inside the caverns of Bruniquel building a series of strange circular structures and using fire. Previously, it had been thought that only modern humans were capable of this type of behaviour or frequented cave systems like this.

confirm this. New analysis at Liang Bua, for example, has found that in the wake of the disappearance of Hobbits, along with giant storks and vultures, by around 50,000 years ago, there was a significant change in the type of stone used to make tools in the site. Chert begins to be overwhelmingly used from around 46,000 years ago, and this has been linked with the sinister presence of modern humans.[2] Whether there was interaction between the Hobbits and us is not yet able to be known with certainty, but after surviving capably on Flores for hundreds of thousands of years, it would seem odd for the Hobbits to disappear prior to our arrival. On the other hand, there is evidence for significant impacts on human populations here from volcanoes. Volcanic ash layers, called tephra, are found on Flores around 47–50,000 years ago. Perhaps Hobbits were significantly affected by these eruptions. We know that further volcanic eruptions around 12–13,000 years ago had major effects on animal populations. Perhaps humans arrived on Flores to find that they were the only people on the island after all and that the previous diminutive occupants had only recently disappeared.

It might just be, however, that the Hobbits survived later. At the time of the initial discovery there was speculation about groups of small people who might have lived on Flores until the last 200 years. Oral accounts suggest that these people, called *ebu gogo* by locals, were living on the margins, in the shadows, near villages, until the time of the initial colonization by the Dutch. Whether this has a basis in fact is not known, for there is no physical evidence. It has been suggested that it is merely a local myth, something that many different cultural groups have, rather like fairies in the forest or witches in the woods. Intriguingly, however, there *are* people of short stature still living on Flores, and close to the Liang Bua site. In a village called Rampasasa there is a population of people who are all under 145cm in height.[3] Scientists have undertaken analyses to see whether they were genetically related to Hobbits. They were not. The genomic data did show, however, that they have genetic variants strongly associated with height phenotypes, rather like Pygmy populations of equatorial Africa. These variants are also found commonly in eastern Asia, but on Flores, amongst this group, they had undergone a strong selection.

This could be related to diet, because a cluster of gene variants linked with fatty-acid production were also highly selected for in the group. This shows that dwarfing has occurred on Flores not once, but twice, and that the Rampasasa people offer no support for an unexpectedly late survival of Hobbits.

It is commonly thought that the main cause of extinctions, both of other humans and many animals, is the rapid arrival of *Homo sapiens*. But how much of this is based on aspects of our inferred human superiority? As the last survivors, does this not mean we are naturally the most successful, the winners in Herbert Spencer's 'survival of the fittest'? This was certainly my hunch when I first began to work on the question of Neanderthal disappearance in Europe in the early 2000s. I thought that Neanderthals probably disappeared quickly following the arrival of the first modern humans into the region. We have already seen that there was strong debate about the degree of overlap and possible evidence for 'acculturation' of Neanderthals by early *Homo sapiens*. At the heart of this was an interpretation of the radiocarbon evidence, as well as the stratigraphic sequences at many sites across Europe.

The best record for Neanderthals, and therefore their disappearance, is in western Europe. When archaeologists examined the sequence of cultural layers in archaeological sites in Europe, they noticed that in some instances there was a thin sterile level between the latest Neanderthal levels – usually the Mousterian or, in some areas, the Châtelperronian – and the earliest modern human level, often associated with the Aurignacian around the period spanning 35–42,000 years ago. This evidence was held to suggest that perhaps there was little or no contact at all because Neanderthals had already disappeared prior to our arrival. Arguments that in some instances there was an 'interstratification' of Aurignacian and Mousterian/ Châtelperronian industries were bitterly debated and, without exception, found to be lacking in solid evidence.[4]

There was a widespread perception in the mid-2000s, however, that in some regions Neanderthals had in fact survived quite late. There were radiocarbon dates on Neanderthal bones as young as 28– 30,000 years in some parts of Europe, for instance.

The south of Iberia was thought to represent a kind of last stand for the Neanderthals, and one of the key locations in this discussion is Gibraltar. Gibraltar is a remarkable place: the 6km wide limestone rock on which it sits seems to erupt from the surrounds of mainland Spain like a seismic blast. Indeed, this is what happened to create Gibraltar, albeit millions of years ago, from the slow collision of the African and European continental plates, which forced the Rock upwards to its present height 436m above sea level.[5]

There are several archaeological sites on Gibraltar that have revealed successive occupations by Neanderthals, and, later, by *Homo sapiens*, the most important of which is Gorham's Cave. The site has been excavated several times over the years, but between 1999 and 2005 Gibraltar Museum Director Clive Finlayson led a new campaign in the deep interior of the cave. (Previous excavations, in contrast, had been at the entrance.)[6] Four archaeological levels were found, the earliest of which was Neanderthal Mousterian.

Radiocarbon dates for this level were surprisingly young. They spanned ten millennia, between 27,000 and 37,000 years ago.* From a single fireplace the dates covered a rather unexpectedly wide 6,000 years. How could this be? To me this evidenced a likely problem with the dating, but Finlayson argued that the area represented a favoured spot and that people had therefore visited over and over again. It is true that Gibraltar has a range of different habitats in easy reach, including the coast, woodlands, wetlands and estuaries. Neanderthals or other humans had the opportunity of living here year-round, so it is no wonder that the sites themselves have evidence for over 100,000 years of human occupation. Finlayson and his team estimated conservatively that the disappearance of Neanderthals on Gibraltar occurred by only 32,000 years ago.

I was sceptical of the radiocarbon dates and wanted to test them, so in 2008 I visited Gibraltar with my then student Rachel Wood and, with Clive, we took samples from fifty bones. I was really hopeful that this would give us a great shot at figuring out whether the

---

* The radiocarbon dates ranged from 23,000 to 33,000 BP, which in calendar years ranges from 27,000 to 37,000 years ago.

original dates were right. Sadly, every single one had absolutely no remaining collagen; they were undateable. The question of whether the original chronology was right lay beyond us. Until there is better evidence, though, I continue to remain sceptical that Neanderthals were at the site in Gibraltar drastically later than they were at other sites in Europe.

Other researchers also think Iberia is an area where Neanderthals survived later than in other places in Europe. They suggest that between 40,000 and 37,000 years ago, Neanderthals in the south of Spain were effectively isolated from *Homo sapiens* by climate change,[7] and that a biogeographical barrier existed running broadly along the valley of the Ebro river in Spain (see Figure 5 on page 28). Between 40,000 and 38,000 years ago Heinrich Event 4, or H4, hit Europe. (Heinrich events periodically introduce bitterly cold conditions caused by the calving of icebergs from the North American ice sheets into the North Atlantic and the switching off of the central heating provided by warm waters coming from North America to western Europe; see Chapter 10.) Iberian temperatures plummeted, forests shrank and semi-desert conditions spread. The south of Iberia may not generally have been an attractive location at this time and this might be reflected in a low number of archaeological sites. I am sceptical that it was only Neanderthals who could be to the south of this inferred biogeographic barrier, however. Ecological modelling work suggests that Neanderthals and *Homo sapiens* had a similar way of life and adaptation to surviving in the Ice Age. Quite why this should be expected to have changed during this period is not clear to me. I think it is unlikely that a frontier or barrier like this operated.

Much of my work has focused on improving the radiocarbon dating record of this period to calculate a more reliable estimate for Neanderthal disappearance. We developed more reliable methods to chemically pretreat and clean samples of bone for AMS radiocarbon dating, including the ultrafiltration of bone collagen that I introduced in Chapter 9. More recently, my laboratory has worked on further improvements to dating bone. We have perfected a method of taking a single amino acid (hydroxyproline) from the bone collagen. This effectively removes *all* contaminants. We observed that when we re-dated samples that had previously given surprisingly recent results, the dates almost always became older than before when we applied these new methods, which was consistent with the improved removal of residual contamination.

Bones found near Neanderthal remains in Zafarraya, in the south of Spain, for example, turned out to be more than 50,000 years old instead of the 30,000 previously measured.[8] Rachel Wood's doctoral work explored the chronology of many of these southern Iberian sites and others in the north. As we saw before, dating sites which claimed to have late Neanderthal evidence showed that in every case they were affected by contamination and, in fact, much older. Her conclusion was that, while Neanderthals *might* have survived late here, there was no robust evidence to support it. A new measurement of one of the two Neanderthals from the site of Mezmaiskaya in Russia became older by more than 10,000 years.[9] The same pattern was repeated at many sites; in fact, I estimated in 2010 that more than 70 per cent of the previous dates from Neanderthal sites were underestimates of their real age.[10] This was a figure that was quite shocking to me. It shows just how much of our initial data was simply plain wrong and how our interpretations of what really happened in the Palaeolithic have been influenced by poor and unreliable data.

By 2014 I had around 200 new dates from forty sites that I had worked on over the previous decade to help me refine the date of Neanderthal extinction. I used the same Bayesian approach that I outlined in Chapter 9. The final model (pictured in Figure 35) suggested that between 39,000 and 41,000 years ago, the Neanderthals disappeared.[11]

In 2015, interesting new evidence was published that supported this estimate. Back in 2003 I had worked on another human fossil, from a site in the Carpathian Mountains of Romania called Peştera cu Oase, or the Cave with Bones. The site was found in a maze of limestone karst tunnels by speleologists (cavers) in February 2002. They had noticed a modern stream that disappeared into the limestone rock on the side of a large hill, bubbling out lower down on the other side of the mountain. They figured there must be caverns inside to be discovered, but accessing them turned out to be difficult. They

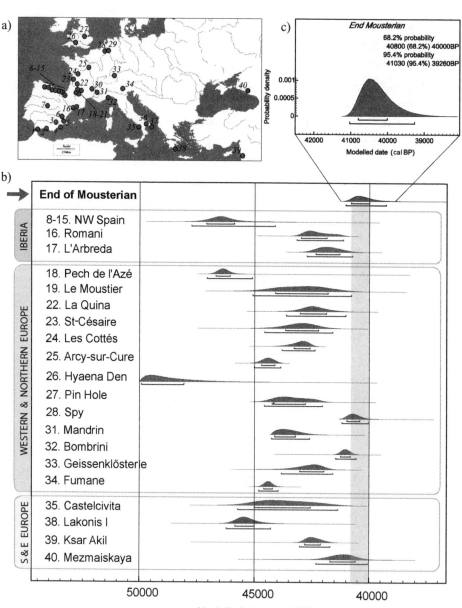

Figure 35 a) Sites with Neanderthal (Mousterian and Châtelperronian) evidence that were dated. b) The Bayesian model of the dated sites. c) The final age estimate for Neanderthal disappearance.

eventually discovered a combination of horizontal caverns and fissures, many of which were filled with water. In one of them they found the bones of Pleistocene animals, mostly bears, littering the floors. One of them noticed the unmistakable form of a perfectly preserved human jaw lying on a bed of calcite.[12] Later, a small piece of that jaw was sent to me for dating. Working with another laboratory, we obtained an age that indicated the specimen dated to between 39,000 and 41,000 years ago. It had a protruding chin, which indicates it is likely to be a modern human. Physical anthropologists working on the specimen, however, identified some of its features as similar to Neanderthals.* The mandible, they said, was interesting in being not particularly modern looking.[13] The shape of the so-called 'mandibular foramen', for example, echoed Neanderthals (this is a small hole in the jaw through which nerves and blood vessels pass). When the molars were measured, they too fell more with Neanderthals than *Homo sapiens*. They concluded that this implied the possibility of some admixture between Neanderthals and modern humans.

More than a decade later DNA was extracted from the Oase jaw. Geneticists at the Max Planck found that 6–9 per cent of Oase's DNA came from Neanderthals. Some of the chunks of Neanderthal DNA were so large that it was clear that a short amount of time had elapsed since introgression had occurred. Somewhere around four to six generations before he lived, the Oase Man had a Neanderthal ancestor – perhaps a great- or great-great-grandparent.† Oase therefore acts as a near-latest possible date for Neanderthal presence in Europe.[14] It also overlaps perfectly with the estimate we had obtained for the last Neanderthals of Europe.

---

* This work was led by Erik Trinkaus of Washington University in St Louis.

† We cannot be more precise than this, because of the way in which our DNA is inherited from successively older generations. We obtain on average 50 per cent of our DNA from each of our parents, but from our grandparents we can receive slightly different proportions: between 23 and 27 per cent. From our great-grandparents, similarly different potential proportions: 9.3 to 14 per cent, and again for great-great-grandparents. So, for this reason, we can only give an estimate of four to six generations when estimating the ancestry that equates to 6–9 per cent Neanderthal DNA.

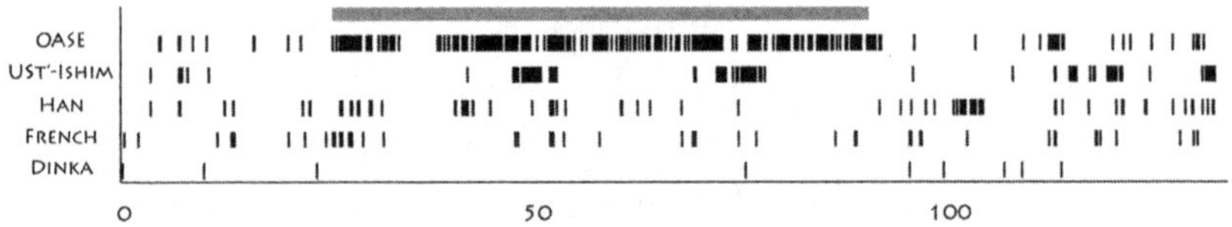

Figure 36 A section of the genome from Oase Man (top line). Black lines indicate the positions of Neanderthal-like alleles on the genome. Below Oase is Ust'-Ishim and below that three modern human genomes, a Han Chinese person, a French person and a Dinka person (from South Sudan), are shown for comparison. The grey shading at the top indicates the area of strong Neanderthal ancestry in Oase, indicating a very recent Neanderthal ancestor. This section of the genome is Chromosome 12, one of 22 chromosomes that were studied (modified after Fu et al. 2015; see note 14).

At the same time as we have been resolving the ages of the last Neanderthals, the dates for the earliest *Homo sapiens* in Europe have also been changing. We have had hints for a while that the earliest European *Homo sapiens* were not the Aurignacians, whom we date after 42,000 years ago. We have evidence of early humans in Italy and Greece dating as far back as 45,000 years old, if not a little more (the oldest levels are still undated).[15] This was based on a new analysis of two teeth from the southern Italian site of Grotta del Cavallo, which were previously thought to be Neanderthal. The tooth re-analysis showed that they were, in fact, modern human and this suggested a pre-Aurignacian presence of humans in southern Europe associated with a transitional stone-tool industry called the Uluzzian. These early dates have been confirmed at a site in Bulgaria called Bacho Kiro, where we find Initial Upper Palaeolithic tools and human bones that date to the same period as Cavallo.[16] Earlier in this book we encountered new evidence from the Grotte Mandrin in southern France, which appears to show an even earlier presence of modern humans, by 50,000 years ago, with tools of an Initial Upper Palaeolithic type that are completely different from those made by European Neanderthals. If we compare the estimates for the latest Neanderthals and the earliest *Homo sapiens* now side-by-side we can reveal a significant overlap in time: up to 10,000 years. This is a much longer span than I ever thought possible. As I said previously, I expected a shorter period of overlap than this. How can we explain it?

I do not think that we should not imagine both groups living side-by-side. Across the continent there were no doubt different human populations over the long duration, rather like a mosaic. In some regions, Neanderthals seem to be present later, around 40,000 years ago, while in others they had seemingly disappeared up to 5,000 years earlier. We can imagine this mosaic as one of broad contemporaneity in different parts of Europe but with a degree of separation being maintained between the different groups. We have direct evidence of overlap and interaction at Oase, of course, but amongst late Neanderthals we have found no evidence yet for the introgression of modern human DNA.[17]* It seems clear that the idea of a modern human invasion of Europe, resulting in a rapid disappearance of Neanderthals, is simply not supported. For over ten millennia prior to their final disappearance 39,000 years ago Neanderthals were our neighbours, and perhaps our friends, maybe occasionally our lovers and sometimes our enemies.

The idea that Neanderthals did not succumb rapidly to modern human advancement into their territories overturns a large amount of published literature, which had suggested the opposite.[18] The model of Neanderthals rapidly disappearing as our own species swept into Europe, superior cognitively and technologically, can be set to one side. That model was built, necessarily, on the basis of a coarse chronology and influenced by the archaeological evidence associated with the European Early Upper Palaeolithic, in which we see a proliferation of symbolic artefacts appearing relatively suddenly in archaeological terms. This suddenness was taken to evidence the relatively rapid arrival and dispersal of modern humans and the consequent demise of Neanderthals. Now, of course, we have evidence that Neanderthals too were capable of creating some of these very objects. Perhaps not in the same number as later Aurignacian *Homo sapiens*, but nonetheless they were capable, and this is attested at

---

* It is possible that some of the late Neanderthals that have been analysed might date earlier than current dates imply and therefore come from the period prior to modern humans being in Europe.

the one or two late Neanderthal Châtelperronian sites that we described in Chapter 3. The arrival of modern humans seems to have been a more complex process, one that lasted for some time.

Could it be that any advantage that modern humans had arose over a longer period of time, during which they outcompeted Neanderthals more subtly? Research has shown that in terms of ecological niches, and the types of environment that moderns and Neanderthals occupied, there were many similarities between the two. Where differences are observed in the types of fauna that Neanderthals or modern humans hunted, I think that climate and environmental influences probably played a more significant role than innate cognitive ability or differences in behaviour. That said, over the longer duration of overlap it is possible that small advantages for one group could have become manifestly larger. Our challenge is to work out what these might have been from a patchy archaeological record.

Periodically our world is dominated by the fallout of a natural disaster: an earthquake, a tsunami, or, increasingly, a climate-related catastrophe. In the academic world this is often accompanied by questions of the effects of similar events in the deeper past. I remember the devastating tsunami that hit northern New Guinea in 1998 spurring research into detecting ancient tsunamis on coastal sites in different places. Volcanic events too periodically remind us of the devastation that they can visit upon both the natural world and humankind. The eruption of Tambora in 1815 resulted in the absence of a summer in the following year. Could a large volcanic eruption have hastened Neanderthal disappearance, or even caused their extinction?

Past eruptions leave evidence in the form of volcanic ash, or tephra, in different archives on Earth, in peat bogs, lakes and marine sediments. If we take a core in a peat bog in any European location the chance is high that we will identify thin layers of ash within the darker peat. Sometimes the volcanic ash is evidenced only by one or two single shards of volcanic glass (microtephra) that can be detected by high-powered microscopes. By dating the peat above and below the layers we can date that past volcanic event. By probing the geochemistry of the volcanic glass, and its minor elements, we can identify the fingerprint signature of a particular volcano and link the ash in the site to its source volcano. In this way we can build up a history of various volcanoes and when they erupted over thousands of years.

Around 39,300 years ago, just as the H4 event was starting to unleash the devastation of full glacial conditions to Europe, a supervolcano erupted near what is now Naples. Vast quantities of volcanic ash erupted from the Campi Flegrei or 'burning fields', blanketing a vast area, stretching from southern Italy north-east towards Russia, Turkey, parts of central Europe and Ukraine with ash and rock.[19]★ We can get an idea of the dreadful impact of this eruption at several archaeological sites, whose occupation was catastrophically ended by it. Kostenki, in western Russia, revealed archaeological living floors that were sealed by the ash.[20] The excavators called it a 'Palaeolithic Pompeii' to describe the likely impact of the eruption. Kostenki is almost 3,000km from the source of the eruption. How must it have affected people living closer to the volcano? Fifty kilometres to the east of the Campi Flegrei lies the archaeological site of Serino. It was buried beneath more than three metres of volcanic ash. The short-term effects were obviously devastating, but it is a little more difficult to assess the long-term impacts. Volcanic ash with a volume of between 250 and 300km³ was released, perhaps producing an effect akin to a nuclear winter[21] and perhaps, like Tambora, affecting climate the following year or longer. The effects would have been relatively brief, though; other work has demonstrated that the impacts of a major volcanic eruption in terms of climate cooling and acid deposition probably only last one or two years at most.[22]

In assessing the impact of the volcano on Neanderthals, however, it is important to note that the people living at Kostenki and Serino were not Neanderthals; they were modern humans. As we have seen, *Homo sapiens* were spread across Europe 40–45,000 years ago, and earlier in the case of Mandrin, and had virtually replaced Neanderthals by the time of the eruption. There is little doubt that, depending on location, human populations were affected, but it seems most likely

★ We call this the Campanian Ignimbrite.

Figure 37 The Serino site. White arrow indicates an Aurignacian (early human) fireplace. Above is volcanic ash from the Campi Flegrei eruption.

to me that Neanderthals did not disappear because of a volcanic eruption. Modern humans, on the other hand, might have been affected. The genomic evidence we have from pre-40,000-year-old modern humans such as that found at Oase and Ust'-Ishim shows that they left no descendants in modern populations. It is only after 40,000 years ago that we see humans with whom we have some genetic ancestry. This suggests there might have been a population turnover around this time. More data are needed before we can blame the Campanian Ignimbrite, but it is possible that, despite its brevity, it was a significant influence on human populations.

Some researchers think disease could have wiped out the Neanderthals, or severely affected them. Transmissible spongiform encephalopathies have been explored as potentially significant.[23] These are a range of nasty prion related diseases, of which Creutzfeldt-Jakob Disease (CJD) is probably the best known. (Prions are abnormally folded proteins.) The CJD variant was linked with the bovine version of the disease, bovine spongiform encephalopathy or BSE, popularly known as mad cow disease, which comes to humans via consumption of

BSE-affected meat. The impact of this type of disease on small, dispersed hunter-gatherer populations was sadly witnessed first-hand amongst the Fore tribe of New Guinea in the mid-twentieth century. There it was known as kuru, and it was spread by funerary cannibalism, particularly the eating of deceased relatives' brains by women and children, as part of a process thought to release the spirit of the dead person.[24] Up to 200 people per year were dying of kuru in the 1950s before it was realized what was causing it.[25] In the case of Neanderthals there is evidence for cannibalistic behaviour, as we have seen, so the transmission of such a disease could have followed a similar pathway to the Fore. It is difficult to conceive of widespread cannibalism continuing across the wide spaces in which Neanderthals lived, however. As we will see later, I think that the Neanderthal population was low and dispersed and this means that the size of individual groups and the networks between them was small as well. In addition, although there are other mechanisms of transmission, such as the shared use of tools carrying the infection during the butchery of animals, this would not be exclusively linked to Neanderthals.[26] I think disease is a potentially important influence at a local level rather than a causative factor in extinction.

Climate change has also been invoked as significant. Some researchers have suggested that solar radiation might have influenced Neanderthal disappearance. Around 42,000 years ago, the geomagnetic shielding on Earth that protects us from heightened solar radiation reduced substantially. This is called the Laschamp event and the decline in shielding from the sun may have been linked to a rise in UV radiation, causing increases in skin cancers, damage to cellular immunity and eye tissue damage.[27] But surely *Homo sapiens* living at the same time as Neanderthals would have suffered too. Researchers involved in this hypothesis suggest that skin colour played a role and that Neanderthals were probably paler and therefore more affected than *Homo sapiens*. As we shall see in the next chapter, however, there are several alleles that play a role in the colour of someone's skin and it is more than likely that, like us, Neanderthals had a range of skin types depending on their location and its climate. Once more, then, I find this scenario interesting but unlikely to be the causal factor.

The major climatic oscillations that occurred while Neanderthals were living in Europe show that they were able to survive significant environmental stresses. It would seem odd to me then, if we are correct that their disappearance occurred at 39–41,000 years ago, that this corresponds to one of the more benign parts of the last glacial. That said, some have suggested that the Neanderthals were a population under pressure from the ongoing harsh climates they faced in northern Europe. It is also true that animals and human hunter-gatherers in cooler, temperate climes have a lower population density compared with people living in more benign regions. The population size of the Neanderthals might be an important factor in exploring why they disappeared.

But how can we explore past population sizes? One approach has been to look at sites and site density, and infer population size tentatively from those,[28] but these methods have been criticized as unreliable.[29] It turns out that the high-coverage genomes we now have provide some informative clues about population sizes and what might have happened to the Neanderthals, and perhaps to the Denisovans too.

Nuclear genomes contain a record of population history going back generations, which can be revealed by statistical analysis.[30] The genome of Denisova 5, the so-called Altai Neanderthal, was particularly informative, because it contained long runs of 'homozygosity'.[31] This is when a person inherits identical alleles on both pairs of chromosomes.* When this is observed frequently across the genome it indicates that two closely related people were a child's parents. The long sections of homozygosity on the Altai Neanderthal genome can only be explained by her parents being very close relatives: first cousins, grandfather–granddaughter, half-siblings or uncle–niece. The degree of homozygosity is much higher than it is for any living group of humans today, and higher than in most organisms on Earth. Tests showed that this was not due to a bottleneck, as occurs when a

---

* These cases are usually denoted by two capital letters (X X) for dominant alleles and two lower case letters (xx) for recessive alleles.

small group becomes isolated with necessarily lower genetic diversity; it was a longer-term problem of small population size.★

In contrast, Denisovans have only a proportion of the heterozygosity of modern hunter-gatherer groups in Africa and South America (20–40 per cent), but lack the long runs of homozygosity we see in the available Neanderthal genomes. The most likely explanation for Denisovans is low genetic diversity in their population rather than inbreeding due to low population numbers.[32]

This information can also be used to reconstruct a demographic history.[33] What is fascinating when this analysis is done is that *Homo sapiens*, Denisovan and Neanderthal population histories show similar patterns to one another, with declining numbers prior to 1 million years ago, but after this, *Homo sapiens* numbers track upwards, while the two other groups decline steadily between 100,000 and 50,000 years ago towards their eventual disappearance.[34]

The high-coverage genomes of *Homo sapiens*, in contrast, show that these groups were not nearly as inbred as the Altai Neanderthals. The parents of Ust'-Ishim Man, for example, were completely unrelated to one another.[35] At the Russian site of Sungir, where three 35,000-year-old *Homo sapiens* were found buried amidst great richness,† analysis

---

★ A Neanderthal from the site of Chagyrskaya, in the Altai, also had significant homozygosity. Almost 13 per cent of its genome was homozygous, suggesting that the person lived within a small group of no more than sixty individuals (see Mafessoni et al. 2020 at note 32). It appears likely that in the Altai, the Neanderthal population was particularly small. This might be because a smaller founding population moved there from the west. Perhaps this is why they failed to maintain their presence there. According to data from Chagyrskaya, Neanderthals disappeared from there around 60,000 years ago. Another high-coverage Neanderthal genome, this time from Vindija in Croatia, also contained low heterozygosity, although it did not have such high runs of homozygosity, suggesting that it was not as inbred as the Altai Neanderthal (see Chapter 6, note 9).

† The three principal burials contained a range of ornaments, beads and pierced animal teeth. The Sungir I body contained around 3,000 mammoth ivory beads, probably sewn into clothing. Sungir II had animal shaped pendants, ivory arm bands, a large mammoth carving and was covered with around 5,000 beads made from mammoth ivory and around 250 Arctic fox canines. Sungir III was similarly decorated. Around 5,400 ivory beads were found on and near the skeleton (see note 36).

revealed that their degree of inbreeding was also low. They were a highly diverse genetic group with an estimated population size that was equivalent to most modern hunter-gatherer groups in places like Amazonia.[36] This suggests that bands such as those of Ust'-Ishim and Sungir maintained strong networks of interaction with other groups and exchanged mates with them.

We know from the genomic data, then, that modern human populations have been larger than Denisovan and Neanderthal populations for long periods.[37] Non-*Homo sapiens* groups come from populations with lower numbers and less genetic diversity. Small populations dispersed in clans or groups across large spaces that experience the unexpected loss of part of the group through bad luck or disease are surely at a heightened risk of disappearance. Archaeological evidence tends to suggest that group and network sizes for Neanderthals and modern humans were subtly different, and bigger and more wide-ranging in the case of moderns.[38] Groups that become isolated from one another or fail to maintain social networks inevitably lose yet more genetic diversity and cultural flexibility, respectively. There may be demographic factors at play in understanding the disappearance of Neanderthals and Denisovans.[39]

It is interesting to note also how recent modelling work has shown that even repeated, small migrations of *Homo sapiens* into Europe over time from Africa might have been enough to tip the balance against Neanderthals. Models assuming no selective advantage for either group, nor any difference in the relative population size, have been shown to predict Neanderthal disappearance.[40] In addition, they predict replacement along approximately the same period of time as we have seen estimated from the archaeological record. Under a so-called model of 'neutral drift', which assumes that there is a random reshuffling of alleles in every generation and no selection acting, Neanderthals might have just disappeared under the weight of small numbers of modern human arrivals. It may have been as simple as that.

We have seen a range of explanations put forward for Neanderthal disappearance, the best case we have for the extinction of one of our close evolutionary cousins. The data concerning Denisovan disappearance and that of other groups is much scarcer, and again we must be cautious in linking cause to effect. My hunch is that the broad story outlined for Neanderthals was similar in other places with different groups, but this remains to be proven.

Although we must be cautious in the light of the incomplete nature of the archaeological record, recent work suggests that differences in population numbers might be significant. Rather than invoke a superiority argument on the part of *Homo sapiens*, it might well be that Neanderthals were inevitably going to disappear because they were fewer in number and less genetically diverse. The inferred cultural superiority of modern humans may be something that happened only towards the end of a prolonged period of overlap of up to 10,000 years, during which both groups interacted and influenced one another. It seems to me that the replacement of Neanderthals probably occurred at different rates and in different places over thousands of years. I also wonder about the possibility that Neanderthals were assimilated into modern human groups through time.

Gradually, it seems, Neanderthals disappeared from the Earth, leaving only the occasional vestiges of their daily lives in caves and open-air sites across Eurasia, as well as some of their genes, which live on in us today. It is this legacy that we examine next.

16.   Our Genetic Legacy

To know ourselves and to work out where we came from, we must know our ancestry. We are now able to probe the genomes of more and more living humans and compare them against the increasing trickle of new Neanderthal and Denisovan genomes to explore our most ancient genetic history. What did we inherit genetically? Did the genes in us today bring positive benefits, or were there negative implications too?

We are at the very beginning of knowing which of our genes derive from our ancient cousins and what their function was. High-quality medical records, including high-coverage genomes from patients who agree to share their genetic and personal data, form the basis for scientific interrogation of large public-health databases.[1] Through this work we can begin to link genes to phenotypes and so place our ancient genetic legacy in context.

I remember well one of the key dates at the start of this new era of genetic archaeology and analysis. In June 2013 David Reich gave a lecture on his recent research to a packed audience at the Oxford Centre for Human Genetics. (I had to cycle up the steep Headington Hill to reach the venue, which shows how keen I was to attend.) Reich had been a doctoral student in Zoology at Oxford University and he gave a fascinating insight into the DNA of Neanderthals and Denisovans. Later that day he and his host, Professor Sir Peter Donnelly, had a very important conversation. Donnelly was working at the time with the UK Biobank (UKB) project, a huge research initiative aiming to use genetics to tackle a wide range of illnesses, including cancer, diabetes, arthritis and depression. The UKB studies a huge group of around 500,000 people, ranging in age from forty to sixty-nine. It is one of the most valuable studies of human health in the world because of the range and breadth of the data that it collects.[2] As well as DNA results, samples of blood, urine and saliva have been taken, along with detailed information about the daily lives of the subjects involved, sourced from extensive questionnaires and clinical evaluations. Through this, we know about all aspects of the lives of the UKB participants, from their diet, disease history, their response to sunburn, the average length of time they use a mobile phone, aspects of their mental health, sexual history, insomnia, visual acuity, skin tone and more. This means that information linking the genomes of the people to their phenotypes, the ways in which these genes are expressed in the environment, can be explored. Somewhat brilliantly, the study group have agreed to make all of their data available to scientists, so many hundreds of phenotypes are available to be compared and more are being added all the time.

Donnelly had been working on designing chips to screen and identify genes that he was interested in exploring within this dataset. He asked Reich if he would be interested in adding some Neanderthal-specific alleles to a custom-built chip so that they could search the UKB dataset for them. Reich was extremely excited about the opportunity to be involved, and later he sent information about 6,000 Neanderthal alleles to Donnelly to include in the analysis.

The first published analyses★ found interesting and statistically significant patterns in the relationship between Neanderthal-derived alleles and the phenotypes in the UKB.[3]

More than half of the traits associated with Neanderthal DNA that initially registered as significant were related to hair and skin biology. Previous studies had obtained evidence purporting to link Neanderthals with red hair,[4] but no link between Neanderthal DNA and red hair could be found in the UKB genome data, suggesting that perhaps it was rare or at very low frequency amongst Neanderthals. They also found that Neanderthal DNA was found in over 60 per cent of people near a gene called *BNC2*, and that individuals in the UKB who carried this Neanderthal DNA

★ This work was undertaken by Michael Dannemann and Janet Kelso at the Max Planck in Leipzig.

reported high incidences of childhood sunburn and poor tanning ability typical of those with fair skin. Interestingly, other segments of Neanderthal DNA in the UKB individuals were associated with more olive skin tones, suggesting that there may have been variation amongst the wider Neanderthal population in skin and hair colour. Given the high number of archaic alleles associated with skin and hair colour, perhaps our *Homo sapiens* ancestors benefited from aspects of these Neanderthal adapted traits and it was advantageous for them after genetic introgression with Neanderthals had occurred.

Do you know people who describe themselves as a 'morning person'? Perhaps you are more of an 'evening person', or perhaps something in between. We call these categories 'chronotypes' and, it turns out, these have a strong genetic basis. There are four categories in the UKB datasets: 'definitely an evening person', 'more an evening than a morning person', 'more a morning person than an evening person' and 'definitely a morning person'. The UKB comparison revealed that Neanderthal alleles associated with genes *ASB1* and *EXOC6* are strongly associated with these preferences. What is interesting is that there is a significant association between latitude and these chronotypes – the further you are from the equator the greater the chance you have the Neanderthal allele at *ASB1*. Your preference for being a morning person appears to be increased by having the Neanderthal allele variant.[5]★

Chronotype is linked with daylight exposure. Given that Neanderthals had adapted and lived for more than 200,000 years in the northern parts of Europe and Eurasia, it seems reasonable to expect adaptation to lower UV levels and sunlight duration in them compared with modern humans ultimately coming out of Africa. There is physical evidence to support this as well. If we plot the size of the eye orbit and eye volume of Neanderthals and compare them with contemporary modern humans living in Africa and lower latitudes,

★ I am definitely a morning person. My DNA, as undertaken by the US biotechnology company 23andMe, suggests this too, with a predicted average wake-up time of 7:05 a.m. Most of this book was written between 5.30 and 8.30 a.m., a period when I am most wide awake and alert.

we find a significant difference. Neanderthals had larger eye sockets, probably to compensate for lower light levels in northern latitudes and the long periods of winter darkness.[6]

There were several other phenotypes detected in the Biobank data which were also significantly linked with archaic alleles. One of these was mood. Another was the frequency of being unenthusiastic or disinterested in the past two weeks. A third, loneliness and feelings of isolation. I wonder whether, like chronotypes, hair and skin tone, the degree of light exposure is a key factor in these mood-related behaviours, hence the Neanderthal link.

Neanderthal genetic introgression may also have negative impacts on modern human carriers today. Research has shown that genetic variants linked to lupus, biliary cirrhosis and Crohn's disease are derived from Neanderthals too.[7] One of the risk haplotypes for Type II diabetes was also admixed from Neanderthals.[8] People who carry the higher-risk mutation are 25 per cent more likely to have Type 2 diabetes compared with those who do not. This does not mean that Neanderthals had a higher tendency to have diabetes, for these genes may function differently in our modern genomes and modern environment. We call this 'exaptation', referring to functions acquired (in this case genetically) for which they were not originally adapted or selected. In the past it may well have been advantageous to have the ability to store sugars – in periods of nutritional stress or starvation, for example. This would have been adaptively advantageous, but now it is the reverse because in the West we often have an oversupply of sugars in our diet.

Initially it seemed that selection over time reduced our Neanderthal ancestry from higher levels close to the date of introgression down to the 1.1–2.5 per cent we have today.[9] Recent work, however, has suggested that this reduction in fact happened quite rapidly, within ten to twenty generations of interbreeding, and that following this the proportion of Neanderthal DNA in non-African populations has stayed at a very similar level.[10]

Can we see the effects of Neanderthal introgression and admixture in living people today? I know I am probably not the only person to have wondered whether someone I know or have seen might be blessed with a tiny bit more Neanderthal than others, but is this really possible or is it simply wishful thinking?

Recent palaeoanthropological studies have attempted to explore this question by combining genetics with high-resolution neuro-imaging scans of living human skulls to study differences in endocranial shape.* We know already that the skulls of Neanderthals and modern humans are similar in size, but have different shapes. Ours are more globular and bulge at the back and towards the front, while those of Neanderthals are more elongated and flatter. A large study of 4,468 living Europeans showed that variants introduced from Neanderthals near two genes did indeed have a subtle influence on the globular shape of their skulls.[11] Intriguingly, the genes implicated are involved in a part of the brain called the deep ganglia and may influence aspects of neural development regions of the brain involved in memory, planning and movement, and perhaps even speech and language too. We are only at the start of probing the functional aspects of genes and understanding what they do and how they work. In future, using gene-editing techniques, it might be possible to study the effects of the presence and absence of certain alleles in stem-cell cultures to work out what they do.

In contrast to Neanderthals, the UKB at the time of writing includes almost no one with introgressed Denisovan ancestry. What we really need is detailed modern human genome and associated phenotype data from people living in places like New Guinea, Melanesia, the Philippines and Australia, populations in which we know there is a proportion of Denisovan DNA. Only when this is done will we truly be able to find out exactly what Denisovan DNA has contributed functionally to people who have inherited it.

Nevertheless, there has been a surge in the generation of new genetic data from diverse populations over recent years, which is slowly enabling more exploratory work to be undertaken on aspects of our inherited Denisovan genetic legacy. In 2016, 300 new genomes from 142 different populations were reported by the Simons Genome Diversity Project (SGDP), a project funded by the Simons Foundation, a US

---

* The inferred shape of the brain as reconstructed from the inside of the skull.

charitable trust.[12] Previously unidentified human DNA sequences highlighted the incomplete nature of the breadth of our genetic diversity. New, never-before-seen DNA sequences from the Khoe-San of southern Africa and from New Guineans, for example, added 11 per cent and 5 per cent respectively to the database of human genetic variation. This allowed a more detailed world map of the proportions of Neanderthal and Denisovan DNA in living humans to be constructed, showing that some groups of people living in South Asia have Denisovan ancestry. This was discovered only because of new genetic data being obtained from people in the sub-continent. Interestingly, the SGDP dataset showed again that people in the east of Asia have a higher proportion of Neanderthal DNA than those in the west. This is the opposite of what we would expect, of course, given that we know Neanderthals were mainly concentrated in the west of Eurasia.

Why would eastern Asians inherit slightly more Neanderthal DNA? In 2020 the answer to this conundrum came from new research on modern human DNA in Africa, which takes account of the fact that there is a small amount of Neanderthal DNA in many living Africans derived from the movement of West Eurasian people with Neanderthal DNA back into the continent.[13] When this DNA is taken into account and subtracted from the analysis, the proportion of Neanderthal DNA in the east and west of Eurasia is much more similar. Once again, more genetic data from previously unsampled populations sheds a brighter light on our ancient past.

We have already seen that modern humans expanded through and into a range of diverse environments on their out-of-Africa journey. This included tropical and rainforest environments, as we saw at Niah Cave and other sites in Southeast and South Asia. Today these are challenging environments for the outsider. When I went to Niah to work I had to ensure that I was properly vaccinated against typhoid, cholera, Japanese encephalitis and, of course, malaria. Without this help I would be reliant solely on my immune system to protect me. The response of our immune systems to parasitical, bacterial and viral attack is crucial to our survival as a species, and the genes that function to protect us are the result of evolutionary processes that have accumulated over millennia. Could the introgression of advantageous alleles from our archaic cousins have also helped in building up our own immune systems? After all, if we are correct about there being a population of Denisovans living to the east of Wallace's Line, then we might expect that they had had time to build up immune responses to some of the diseases that lurk there.

It turns out that the answer is an emphatic 'yes'. Genes associated with immunity appear to be amongst the most positively selected genes introgressed to modern humans from our archaic cousins. And there are a lot of them: around 400 gene variants inherited from Denisovans are concerned with either immunity or diet. What do they do? We are at the very start of working this out, but some interesting data are beginning to be published.

One of the variants with the highest frequency in living Papuans so far identified is called *TNFAIP3*. *TNFAIP3* codes an immune controller protein called A20. Natural polymorphisms (genetic variants) of this gene are associated with overactive immunity in autoimmune conditions such as arthritis, multiple sclerosis, inflammatory bowel disease and psoriasis. Researchers have identified that some variants of this gene increase the inflammatory response, which might be how variants of *TNFAIP3* associate with autoimmune disease. In one case, a gene variant of *TNFAIP3* is one of these Denisovan introgressed genes.[14] The specific Denisovan variant of the gene (called *I207L*) involves a change in a single amino acid in the A20 protein (isoleucine to leucine), and this is enough to produce a stronger inflammatory response in people with this variant compared to people without it. To test the autoimmune response of this particular gene, researchers took the variant and introduced it to mice, which they infected with a virus called the Coxsackie virus strain, a highly contagious hand, foot and mouth virus that strikes children under the age of five. Children go to bed at night feeling fine but wake in the morning with high temperatures and covered in bright red blisters. The mice with the *I207L* variant had significantly stronger immune reactions and resistance to the virus than those without it.

The distribution of this variant in modern people is striking:

people east of Wallace's Line have an extremely high frequency while those to the west have virtually none. *TNFAIP3* may have been positively selected in humans after introgression from Denisovans because of the added protection it conferred to unusual tropical microbes. When immune cells from people with the *I207L* variant were tested they showed increased immune reactions. Increased immune reactivity may act to protect people from agents of disease such as microbes. In Island Southeast Asia and Oceania, between 25 and 75 per cent of people are carriers. Elsewhere in the world, there are none. Interestingly, the *I207L* variant was also found in the Denisovan girl from the Altai, Denisova 3, suggesting that it must date prior to the separation of the various Denisovan populations in East Asia and after the split from Neanderthals, since the Neanderthal genomes from Denisova Cave do not have it.

A recent study of DNA from Near Oceanian people showed several important structural deletions in genes that are central to detecting viruses and regulating the ability of the body to fire antiviral immune responses.[15] Increasingly, one begins to realize that these genetic differences will be of great importance to modern medicine. On the one hand this knowledge enables us to understand more about how different groups have developed the ability to fight diseases, and on the other it will help us in future to provide targeted medical care to people from different backgrounds.

The introgression of Denisovan DNA has also been linked with understanding how it is that Tibetans are adapted to coping with life at high altitude, where there is 40 per cent less air than at sea level. Some of these adaptations concern newborn babies, who weigh less on average than babies at lower altitudes. For every 1,000m of increasing altitude, newborn weights reduce by, on average, 100g. They also have lower levels of haemoglobin in their blood[16]★ with higher

★ Lower levels of haemoglobin in the blood seems strange, because when people not habituated to living at altitude go to higher elevations, they acclimatize by slowly increasing the level of blood haemoglobin. In the case of Tibetans, however, they have lower levels probably because increasing haemoglobin levels carry associated risks due to elevated blood viscosity and cardiac problems, including possible cardiac arrest. See note 20.

arterial oxygen saturation at birth and early life than Han Chinese babies living at the same altitude.[17] This is thought to be a response to hypoxia[18] – when a part of the body fails to get enough oxygen to operate properly. This suggests that there has probably been selection going on in Tibetans for genes that enable them to cope better with higher altitudes.

The key gene in this case is called *EPAS1*. All human populations carry *EPAS1*, but in the case of Tibetans the gene has a specific variant that produces a protein that provides the bearer with a hypoxia response and the ability to live more readily at altitude.[19]

To explore the genetic aspects of this adaptation a team sequenced the exomes of a group of Tibetan villagers.* They focused on the specific section of the *EPAS1* gene and found that SNPs in the Tibetan version of the gene had a distinct pattern when compared against a group of Han Chinese people and a group of Danes, both, of course, habituated to much less rarefied air.

But how did Tibetans obtain this particular gene variant? Was it the result of mutation or was it derived from somewhere else? The close genetic similarity between Han and Tibetans made it a very tricky problem to solve. Despite their close genetic relationship, *EPAS1* has a 78 per cent difference in its expression between Tibetans and Han Chinese, so it is clearly important for people at altitude. Despite the evidently strong selection for the gene it was impossible to distinguish the route by which it had become so dominant in the Tibetan genomes.

The team looked into data from the 1000 Genomes Project to see whether the gene sequence in Tibetans might have come from another human population, but nothing like the Tibetan version of the *EPAS1* gene was evident in that entire dataset except in two Chinese people. Strange, because once again Han Chinese are very closely related to Tibetans genetically.

---

* Exomes are the genes that play the key role in directing protein production. The team was led by Emilia Huerta-Sanchez, now at Brown University.

Then they checked against the newly available high-coverage Neanderthal genome. Nothing there either.

There was only one human genome left in the world to check against: the Denisovan genome.

And suddenly there it was – in the high-coverage genome of the Denisova 3 girl . . . an identical set of haplotypes.[20]

The DNA had come to Tibetans from Denisovans. Immediately, it all made sense.[21]

As with other modern humans, interbreeding with Denisovans introduced DNA, including the section surrounding the *EPAS1* gene, at some point in eastern Asia. The Tianyuan Man's genome showed us that this probably happened some time before he lived (you will recall he was dated to 40,000 years ago). Han Chinese and Tibetan people share remarkably similar levels of Denisovan DNA (0.4 per cent), which suggests that it is likely that the introgression from Denisovans came into their *ancestral* population, rather than later into the Tibetans only. When the descendants of those modern humans began to settle in the Tibet plateau, the *EPAS1* variant started to sweep through the population because it was highly advantageous. This is another lovely example of what is called 'adaptive introgression',[22] where populations benefit from advantageous introgressed DNA.

Interestingly, we see the same pattern in the region's animals (which also all have *EPAS1*). The Tibetan mastiff, for example, owes its ability to survive at high altitude on the Tibetan plateau to an introgression from grey wolves that were already living there.[23] Tibetan cattle derived a hypoxia-related gene variant to help them cope with altitude by interbreeding with local yaks, from whom they derive just over 1.2 per cent of their genome.[24] It seems every animal is in on the game of taking advantage of prior adaptation, through genetic admixture. Humans are no exception.

The discovery of *EPAS1* in Tibetans immediately reminds me of the Denisovan jawbone from Baishiya Karst Cave in Xiahe that we met in Chapter 7. It is tempting to wonder whether Denisovans living on the Tibetan plateau might have interbred with humans there, and that perhaps this was where that advantageous variant originally appeared. We know now that modern humans have been living on the Tibetan plateau since as early as 40,000 years ago, based on new evidence from an archaeological site called Nwya Devu.[25] This is 4,600m above sea level and is in the interior of the Tibetan plateau. This tells us that humans have been here at least since these times, and perhaps earlier. Like mastiffs and cattle, perhaps the Tibetan plateau was the place where this beneficial gene variant came into arriving modern human populations.

Denisovan introgression can also help us to explain aspects of the lifestyle and adaptation of the Greenland Inuit. To explore genes linked with adaptation to life in extreme Arctic conditions, researchers explored the genomes of Greenlanders to find genes that had high levels of positive selection in that population. One of the most positively selected signals in the Inuit studied comprised two genes called *WARS2* and *TBX15*. These are linked with body fats and body-fat distribution,[26] in particular to a type of fat called brown fat. This is common in newborn children, where it enables the bearer to generate heat by burning calories. *TBX15* influences the body's reaction to cold climatic conditions. It seems unsurprising that one would see this gene selected positively in an environment like Greenland. Almost 100 per cent of Inuit have these genetic variants compared with other populations. (Interestingly, these genes are highly 'pleiotropic', which means that they are associated with a variety of other types of traits. These include ear shape, stature, hair pigmentation, facial morphology and fat distribution.) The sequences in Greenland Inuit are most closely related to those from the high-coverage genome of the Denisova 3 girl, suggesting strongly that her population is where they originated and introgressed from.

We have learnt substantial amounts from studies that explore and interrogate the genomes of modern people. We have seen that interbreeding has introduced genetic variants that were advantageous to people in certain environments, and are therefore present at high frequency in these populations. We have also seen that other genetic variants derived via Neanderthals have introduced the potential for certain people to have greater susceptibility to illnesses, such as Type II diabetes and lupus. It is clear, then, that there are both positive and negative aspects to our legacy of genetic introgression. What's really exciting is that this genetic detective work is only just beginning. Over the next few years we can expect many more revelations about modern health and how our genes, both derived and ancestral, make us who we are today.

17. The World Before Us

Our understanding of the crucial period of late human evolution, from around 200,000 to 50,000 years ago has transformed fundamentally in little more than a decade. Fresh excavations and the application of new and improved scientific methods, particularly ancient genomics, has shown us that the world before us is far more complex than we could have ever thought possible. In 2010 the Neanderthal Genome Project revealed for the first time that we had interbred with Neanderthals.[1] In the same year, the genome of Denisovans was sequenced. We know that this group consisted of at least two, perhaps three, different populations that appear to have been widely distributed across eastern Asia and the islands of Southeast Asia and Melanesia. Rather like our ancestors, then, Denisovans must have been capable of living in a range of diverse environments and climates across a singularly vast area. We have a great deal more to learn regarding the specifics of how they did this: what technology they had, the stone tools they made and aspects of their daily lives. Denisova Cave itself is the best record available thus far. Here, Denisovans were present from around 200,000 years ago. The archaeological evidence suggests that they lived in a rather similar manner to other Palaeolithic humans. There are hints that late in their history, however, they may have made the ornaments and artefacts that archaeologists infer represent behavioural complexity – traits that for many decades were associated only with our species. The discovery of other potential sites of Denisovans, like Baishiya Karst Cave, raises the possibility of more significant new insights into their archaeology being made over the next few years. Archaeological sites in China that may yield new Denisovan remains also have the potential to reveal a wealth of knowledge about their adaptation. At present, we know much more about Denisovans from their genetics than we do from archaeology. As we have seen, there are just five Denisovan specimens from Denisova Cave, and with the Xiahe mandible, a total of six – although there is the tantalizing possibility of others that currently lack genetic confirmation.

Alongside us (*Homo sapiens*), Denisovans and Neanderthals, we have also met other recently discovered human relatives. The Hobbits of Flores, *Homo luzonensis* on the island of Luzon and *Homo erectus*, who potentially survived as late as 50,000 years ago in Island Southeast Asia. In Africa, there is a distinct possibility of more humans overlapping with the early record of *Homo sapiens* too. We briefly met *Homo naledi* and *Homo heidelbergensis* in Chapter 2, both of which have dates in Africa touching the last 200,000 years[2] – yesterday in geological time. The potential for more hitherto unknown humans is revealed by genetic 'ghost populations' in Africa and Eurasia.

We humans have always thought of ourselves as unique; it turns out that in evolutionary time this uniqueness did not exist until yesterday.

Evidence shows that these different groups of humans regularly interbred when they met, raising questions about how we all ought to be defined taxonomically. There are major implications for our traditional concept of 'species', and what a species is. With respect to the genus *Homo*, this clearly needs to be looked at with fresh eyes, given what we have witnessed in recent years. Speciation is deemed to have occurred when two populations are not capable of producing fertile offspring. Since Linnaeus began the science of taxonomy, we have become used to viewing species as separate and able to be fitted into neat box-like categories. They can't. There are fuzzy boundaries to most species. From coyotes and wolves to bears of all types to mice, frogs, finches, marmosets and more, there is an increasingly long list of species that regularly hybridize with others.

Baboons are an interesting primate example that offers a sobering parallel to the genus *Homo*. Like *Homo*, baboons began to radiate outwards into different regions of Africa from about 2 million years ago.[3] There are six different lineages currently known and their home ranges include vast tracts of the central equatorial zone, southern Africa, eastern parts of Africa and more. Several lineages have ranges that are either close to or overlap with those of other groups. These various baboon lineages look quite different from one another physically and, within their own home ranges, they appear quite homogeneous. A closer look at their genetics, however, reveals unexpected complexity. There is clear evidence for genetic exchange between some lineages and these are most pronounced in so-called 'hybrid zones', which are located on the boundaries of the ranges of different groups. There is also evidence for genetic exchange with extinct lineages and with other living primates (for example, with the kipunji, a monkey living in the highlands of Tanzania[4]). Like us, there is also evidence in baboons for the introgression of DNA from so-called ghost lineages.

Interestingly for us, baboons in hybrid zones also have distinct morphological characteristics that derive from this process of hybridization. Some of this involves increases in size. We call this 'hybrid vigour' (heterosis); it's often seen in animals that have bred with others outside their own species. Ligers, for example, the sterile offspring of a lion and a tiger, are much larger than their cross-species parents. Hybrids are often characterized by larger faces, bigger bodies and dental anomalies.[5]★

It is interesting in this light to remember back to Chapter 7 and the Denisova 4 tooth, a tooth that is strangely comparable in size with much earlier and unrelated hominins. Is this the tell-tale sign of possible recent hybridization? It seemed curious that this Denisovan tooth could be so different and so much larger than a Neanderthal tooth, when they shared a common ancestor some 500,000 years ago.[6] Could hybridization also explain the curious mosaic of features outlined in Chapter 7 that we saw in China, in the human remains excavated from the sites of Xuchang and Xujiayao? Do these specimens reflect gene flow or recent hybridization?

Hybridization may play a role in creating new species too. Variation is the engine room of selection in evolution, and variation as a result of hybridization may be one way that this happens quickly. I recall the curious morphology of *Homo floresiensis*, whose teeth appeared unique and therefore difficult to correlate to anything in the

---

★ Dental anomalies include extra teeth, as well as teeth that are larger, rotated in shape and have anomalous sutures, cusps and lines.

fossil record of *Homo*.[7] We described in Chapter 12 how debate has focused on whether the Hobbit is a dwarfed *Homo erectus* or a slightly less dwarfed *Homo habilis*. An alternative might be that it derives from some kind of hybridization, say between *Homo erectus* and another hominin, perhaps a Denisovan.[8] (Another possibility is that Hobbits and *Homo luzonensis* might in fact be dwarfed Denisovans.)

I have long wondered whether Denisova Cave and the Altai region could be a hybrid zone. If we consider Neanderthals as predominantly west Eurasian and Denisovans as the eastern equivalent, then the Altai region falls in the middle, echoing rather nicely the hybrid zone locations we see between different baboon groups in Africa. There is evidence that Neanderthals must have moved into the Altai from the west of Eurasia on more than one occasion; in fact, it looks as though there were at least two different groups of Neanderthals there at 120,000 and 70,000 years ago.[9] Our ZooMS work and the ancient DNA analysis of bone and sediment show that Neanderthals and Denisovans alternated at the cave several times. New evidence reveals that genetic introgression happened between the two populations on a regular basis.[10] Denny is unusual in being a first-generation hybrid, but every single ancient genome from Denisova Cave has some degree of admixture. Our lineage is the same. Old models of human evolution that comprise isolation and divergence have to be replaced by models that include gene flow, genetic introgression and admixture. The branchy tree of human evolution should instead be replaced with a wide braided river, whose tributaries often flow into and away from one another.

We mentioned in Chapter 3 that when different species interbreed there are sometimes sterility problems for the offspring. We saw that there are tracts of the human genome without any Neanderthal DNA and that this is similar in hybrids that have breeding difficulties. Recent work has shed further light on this. An Oxford team compared the mitochondrial and nuclear DNA sequences of a range of hybridizing pairs of mammals whose offspring possess a range of fertility, from sterile to perfectly fertile.[11] They focused on the cytochrome b gene (*CYTB*) and produced a measure of the genetic distance between pairs of animals and their ability to reproduce with one another. The results showed that there were sharp thresholds in the divergence values, which meant they could use the values to predict degrees of fertility. They were able to show that there would have been no reduction in fertility between humans, Neanderthals, Denisovans, or even with older species of our genus such as *Homo heidelbergensis*. In other words, we ought to have expected the successful interbreeding between these groups even without the evidence provided so spectacularly from nuclear DNA.

As we have seen throughout this book, interbreeding and genetic introgression have conferred a range of benefits to us. Our success in evolutionary terms has been sharpened considerably due to our hybridizing tendency.

I think that this hybridizing nature can also be extended as an analogy to aspects of human behaviour and culture. As we saw in Chapter 3, the idea that *Homo sapiens* are uniquely modern and possess 'behavioural modernity' has been overturned in recent years in the light of compelling evidence that Neanderthals (and perhaps Denisovans too) appear to have had similar abilities and behaviour. In this sense there was no sudden 'human revolution' that modern humans were uniquely responsible for, or, if there was, it was a revolution that other human groups probably contributed to. Evidence from Africa shows that the development of more complex behaviours and the adoption of new technologies were earlier than in Europe and Eurasia and more drawn out, rather than a sudden revolutionary burst.[12] Neanderthals in Europe appear to have been following a similar kind of developmental trajectory.

I wonder whether ideas and innovations may have been generated through the contacts that happened 50,000 years ago, when Neanderthals, Denisovans and modern humans (and others) met one another. Imagine the opportunities inherent in meeting a completely new group of people who were doing things a little bit differently, more creatively, perhaps better than your own group. Humans love to copy and imitate. We see this in more recent archaeological periods, when new trade links introduce exotic items that are then copied and manufactured by locals, sometimes from different materials. The opportunity to learn something new, to make a tough life in the Stone Age more bearable, to capture game without so much difficulty or to decorate oneself in a new way, must have been a very attractive one. The contact period, or periods, between different groups of people may have witnessed an exchange of ideas and innovation driven by this beneficial collision of worlds.[13] Perhaps the creative burst we see in the Early Upper Palaeolithic was the result. Hybridization may therefore have occurred culturally as well as biologically, so to speak, and conferred biological as well as cultural benefits on our ancestors.

We have also seen that the idea of a single group of successful modern humans sweeping out of Africa and rapidly replacing all other groups is unlikely. Ultimately it did happen – we are the only remaining humans on the planet, after all – but the entire process seems to have been much more subtle and drawn out than we previously thought. We now know that there were long periods of overlap between us and Neanderthals in Europe, and probably with Denisovans in other regions too, if the latest genetic evidence in New Guinea is any indication. This overlap was probably not spatial in the sense of close communities living all together; the DNA evidence negates this. It seems to me that the most parsimonious reading of the accumulated evidence thus far is for a mosaic of human groups, geographically separated but meeting and interbreeding occasionally.

Perhaps more work will show that this is a simplification and that different groups of humans sometimes really did inhabit caves and other sites together. And why not? One of the most exciting things about the new sediment DNA research we explored in Chapter 11 is that this is a tool that could very well reveal such behaviour through the recovery of tiny traces of DNA in the detritus of archaeological sediments. I wonder whether there were times, thousands of years ago, when people from different groups found themselves in the same place and spent part of a summer in close proximity before going their separate ways and moving on to somewhere else. Perhaps it was during a time like this that a male or female member of one of the groups decided to pair up with someone from another group and leave with them. Perhaps this is what happened with one of Denny's parents 120,000 years ago in the Altai Mountains.

Mate exchanges like this between groups may have been a rare occurrence, though. We know from genetic evidence that Neanderthals, and perhaps Denisovans too, were low in terms of their population numbers and group sizes at certain times and places. I think this lack of genetic diversity and few links with other groups may have played a part in the demise of these groups, as we saw in Chapter 15.

The role of drastically variable and changing climates in the story of human evolution is also evident. As we witness first-hand the potential for climate disaster in our own lifetimes, we can more easily imagine the catastrophic effects of some of the major climate swings of the last 150,000 years upon small bands of humans spread thinly across the seemingly unending expanses of Eurasia. One particularly crushing winter, one piece of bad luck, and things would have been extraordinarily perilous for small and isolated populations. The death of an experienced member of a group results in the death of ideas and traditions that may then have to be reinvented. In the modern world we see that creativity is more common in denser and more populated places. In fragile worlds with fewer people, ideas and knowledge are precious commodities, and can disappear in an instant.

Finally, in this light it also seems to me that luck too played a role in the story of our early human ancestors. But for a different set of biological or environmental circumstances, I think that it might easily have been a Denisovan or Neanderthal population that survived instead of us and went on to disperse into the rest of the known world like the invasive species we eventually became. The fact that Neanderthals were present on the planet for more than 250,000 years shows that they were a well-adapted and successful experiment in evolution. We have evidence now that they were capable, like us, of wide dispersals and movement, from Western Europe to the Altai Mountains. We may find that Denisovans, and others, were even more successful. But they too ultimately disappeared. The human lineage went through bottlenecks, where our genetic diversity and population fell dangerously low. We could easily have joined these others in extinction but for moments of luck.

When we talk about living and dying, succeeding and failing, however, we must always remember that these human cousins of ours are not completely lost. They live on in us in different and now fragmented parts of our own DNA. They are within and part of us. This is a quite beautiful thought. More than 20 per cent of the Neanderthal genome can be recovered from modern human populations, perhaps more.[14] We will soon find out how much of the Denisovan genome can be sieved out of us and put back together, and discover more and more about what we inherited and what we did not. The stunning realization that we receive from these lost cousins minute traces of genetic coding without which we could not live at altitude in Tibet, or as well in the cold of Greenland, or resist disease in Melanesia, or have adapted so readily to new environments outside our African homeland, is surely one of the great contributions of ancient genetics to understanding our current human condition. The success of our own species and the wide and exuberant diversity of humanity are due in part to these precious genetic gifts. We are not simply human, we are the sum of all of the branches of that braided river that touched and parted on our way to today.

We have learnt much in the last decade, but there remains so much more to find out and many questions that are as yet unanswered. With new excavations and the application of more cutting-edge science, we will surely obtain many more insights into Denisovans, Neanderthals, Hobbits, *Homo luzonensis, Homo erectus* and perhaps other human groups yet to be discovered.

If the last decade is anything to go by, we are in for an exciting ride.

Acknowledgements

There are many people who have helped in the course of writing this book, and for their advice, support and insight I am tremendously grateful.

First, I am indebted to my agent Joanna Swainson at Hardman and Swainson for her guidance, help, comments and support. I am also grateful to the team there, particularly Thérèse Coen.

Connor Brown and Daniel Crewe at Viking/Penguin Random House have been brilliant editors and advisors and I thank them most sincerely. I am very grateful to Trevor Horwood for his insightful and careful copy-editing.

I would like to acknowledge several of my colleagues for their comments and suggestions on parts of the text: Michael Shunkov, Anatoly Derevianko, David Reich, Thomas Sutikna, Johannes Krause, Janet Kelso, Iain Davidson, Matthias Meyer, Elizabeth Veatch, Simon Hillson, Bence Viola, Dongju Zhang, Henry Gee, Samantha Brown, Jesse Dabney, Xinjun Zhang, Chun-Hsiang Chang, Murray Cox, Peter Brown, Jacques Jaubert, Nathan Zammit, Shane Grey, Paige Madison, Stefano Benazzi, Peter Ditchfield, Charles Higham and Katerina Douka. I am very grateful to all of them while acknowledging that any errors remaining are entirely my own.

I acknowledge all of the colleagues and friends who work so hard in the field and the lab to extract the archaeological evidence that forms the basis for the underpinning data and to the hundreds of students and workers on sites year in, year out.

I want to single out some of my more important collaborations while acknowledging that I have almost certainly left people out. Please forgive any omissions.

I thank the entire team of friends and colleagues at Denisova Cave and the wider Altai, particularly Anatoly Derevianko and Michael Shunkov for their collaboration, friendship and support. It is such a great privilege to be a member of the Denisova team. I also want to thank others at the Institute of Archaeology and Ethnography, Russian Academy of Sciences, Siberian Branch in Novosibirsk, particularly Andrey Krivoshapkin, Kseniya Kolobova, Maxim Kozlikin, Alexander Agadzhanian, Evgeny Rybin and Natalya Belousova. Without Vladimir Vaneev and Elena Pankeeva working in the Altai would have been much more difficult. I am grateful to Sergei Zelinsky for his photography and friendship.

At the Max Planck Institute for Evolutionary Anthropology in Leipzig I am indebted to Svante Pääbo and his extraordinarily talented team of researchers and students, in particular Matthias Meyer, Janet Kelso, Viviane Slon, Kay Prüfer, Mateja Hajdinjak, Ben Vernot, Fabrizio Mafessoni, Diyendo Massilani and many others.

Bert Roberts and Zenobia Jacobs (and their team at the University of Wollongong) were fabulous and generous collaborators on the chronometric work at Denisova.

I thank Ludovic Slimak and Laure Metz and their team at Grotte Mandrin, particularly Clément Zanolli, who worked on the key tooth remains from the site. I am grateful to Lludmila Lbova and her team for collaborating with us on the sediment work at the sites of Khotyk, Varvarina Gora and Kammenka.

I am very grateful to João Teixeira, David Reich and Chris Turney for permission to discuss data prior to it being published.

The financial support of the European Research Council, the Natural Environment Research Council, the Leverhulme Trust, Keble College Oxford and the John Fell Fund (Oxford) to my research is gratefully acknowledged.

I want to express my deep thanks to all of the PalaeoChron team: Katerina Douka, Thibaut Devièse, Marine Frouin, Christopher Bronk Ramsey, Dan Comeskey, Jean-Luc Schwenninger, Rachel Hopkins, Samantha Brown, Lorena Becerra Valdivia, Eileen Jacob, Natasha Reynolds, Aditi Dave, Michael Buckley, Monty Ochocki, Maria Emanuela Oddo, James McCullagh, Alexander Benn, Ségolène Vandevelde, Christa Wathen, Cara Kubiak, as well as Rachel Wood. Much of the work described in this book was done in collaboration with them, and by them, and always as a team. I want to thank

Samantha Brown in particular for the amazing time we had working on Denny.

Without the late Roger Jacobi of the British Museum and the Natural History Museum I probably would not have ending up working in the Palaeolithic. His enthusiasm and need to solve tricky problems enthused me too.

I am grateful to all of the team at the Oxford Radiocarbon Accelerator Unit, past and present, for their careful laboratory work, good humour and professionalism. I am lucky to work within the Research Lab for Archaeology and the History of Art (RLAHA) at the School of Archaeology in Oxford, and I thank all my colleagues and students there, past and present, for the coffee-time chats, the sharing of ideas, seminars, discussions and the time that they have given me over the years. It is a privilege to work in such a stimulating and exciting environment.

Professor Greger Larson at the RLAHA has been a constant friend and source of encouragement. I am thankful that there exist generous souls like him and Frances Larson to make science fun.

I thank all of my friends and colleagues at Keble College Oxford, particularly Sir Larry Siedentop for his advice and deep interest in all things human and Sir Jonathan Phillips and Dame Averil Cameron for their support and guidance.

Without the Allan Wilson Centre of New Zealand, and their invitation to me to give lectures across the country in 2015 I don't think I would have written this book, so I am grateful to the New Zealanders who came to the talks, asked questions and, most of all, asked me to recommend a book to them on Denisovans (to which my answer was 'Sorry there isn't one!'). In this light I thank Wendy Newport-Smith and Glenda Lewis for their organization and help on that trip.

The amazing love and support of my large family have been extremely important to me. I would like to thank my father, Professor Charles Higham, and my mother, Polly Higham, my brother James, my sisters Emma and Caroline, and all the wider Higham family in New Zealand and further afield. My step-children Sian Hodgson and Rosie Thomson are thanked for their love and support.

My children Joe, Miriam, Angelo and Elektra put up with long hours of work and their dad being away a great deal.

Finally, I owe more thanks than can possibly be articulated to Dr Katerina Douka. She is my wife, collaborator, friend and advisor, and without her support, criticism, comments, ideas and patience this book would never have come to fruition. Ευχαριστώ κατερίνα μου.

Picture Credits

Figures

Figures 1, 2, 5, 7, 8, 10, 12, 13, 15, 17, 18, 23, 26, 31, 32, 33 and 35 were drawn by the author. I used D-Maps (https://d-maps.com/) to create the base maps used in this book. Figure 3: Philipp Gunz, MPI-EVA Leipzig. Figure 4 is adapted from A. Kruger, P. Randolph-Quinney and M. Elliott 2016. Multimodal Spatial Mapping and Visualisation of Dinaledi Chamber and Rising Star Cave. *South African Journal of Science* 112(5/6). Figures 6, 11, 16, 19, 20: photographs by the author. Figure 9: Siberian Branch, Russian Academy of Science. Figure 14 was redrawn by the author from David Reich, pers. comm. Figures 21, 22, 25 and 29 were drawn by Katerina Douka, MPI-SHH (Jena). Figure 24 was redrawn by the author after Belfer-Cohen and Goring-Morris (2017). Figure 27: photo by Samantha Brown. Figure 28 is redrawn after Racimo et al. (2015). Figure 30: photo by Iain Cartwright. Figure 34 is courtesy of Peter Brown. Figure 36 was redrawn by the author from Fu et al. (2015). Figure 37: photo with permission from A. Ronchitelli, Siena.

Plate sections

See page x for image numbering.

Plates 1, 2, 7, 17, 18, 19: T. Higham, University of Oxford. Plates 3, 13, 14, 16: Siberian Branch, Russian Academy of Sciences, Novosibirsk. Plate 4: Marco Peresani, University of Ferrara. Plate 5: S. Finlayson, Gibraltar National Museum, Gibraltar. Plate 6: Dirk Hoffman, University of Göttingen. Plates 8, 10, 12, 21: MPI for Evolutionary

Anthropology, Leipzig. Plate 9: Frank Vinken for Max Planck Society. Plate 11: Maayan Harel. Plate 15: Photo by Hilde Jensen; copyright, University of Tübingen. Plate 20 (*top* and *left*): Liang Bua Team. Plate 20 (*right*): Peter Schouten/National Geographic Society/ University of Wollongong. Plate 22: Dongju Zhang, Lanzhou University. Plate 23: Photography É. Fabre, SSAC; copyright: Bruniquel Cave.

References and Notes

Chapter 1: Introduction

1 Smith, T. M. et al. 2018. Wintertime Stress, Nursing, and Lead Exposure in Neanderthal Children. *Science Advances*, 31 October 2018: EAAU9483.
2 Ibid.

Chapter 2: Out of Africa

1 Stringer, C. B. and Andrews, P. 1988. Genetic and Fossil Evidence for the Origin of Modern Humans. *Science* 239: 1263–8.
2 Hajdinjak, M. et al. 2018. Reconstructing the Genetic History of Late Neanderthals. *Nature* 555: 652–6.
3 Wolpoff, M. H. et al. 1994. Multiregional Evolution: A World-Wide Source for Modern Human Populations. In M. H. Nitecki and D. V. Nitecki (eds.), *Origins of Anatomically Modern Humans*. Interdisciplinary Contributions to Archaeology, Boston, MA: Springer, pp. 175–99.
4 Cann, R. L., Stoneking, M. and Wilson, A. C. 1987. Mitochondrial DNA and Human Evolution. *Nature* 325: 31–6.
5 Henn, B., Cavalli-Sforza, L. L. and Feldman, M. W. 2012. The Great Human Expansion. *Proceedings of the National Academy of Sciences* 109(44): 17758–64.
6 Linz, B. et al. 2007. An African Origin for the Intimate Association Between Humans and *Helicobacter pylori*. *Nature* 445(7130): 915–18.
7 Scerri, E. M. L. et al. 2018. Did Our Species Evolve in Subdivided Populations Across Africa, and Why Does It Matter? *Trends in Ecology & Evolution* 33(8): 582–94.
8 Hublin, J.-J. et al. 2017. New Fossils from Jebel Irhoud, Morocco and the Pan-African Origin of *Homo sapiens*. *Nature* 546: 289–92.

9 Richter, D., Grün, R., Joannes-Boyau, R. et al. 2017. The Age of the Hominin Fossils from Jebel Irhoud, Morocco, and the Origins of the Middle Stone Age. *Nature* 546: 293–6.

10 McDougall, I., Brown, F. and Fleagle, J. 2005. Stratigraphic Placement and Age of Modern Humans from Kibish, Ethiopia. *Nature* 433: 733–6.

11 White, T., Asfaw, B., DeGusta, D. et al. 2003. Pleistocene *Homo sapiens* from Middle Awash, Ethiopia. *Nature* 423: 742–7.

12 Ibid.

13 Scerri et al. 2018.

14 https://isthmus.com/news/news/anthropology-prof-john-hawks-and-uw-madison-students-dig-up-crucial-remnants-of-early-hominids/ Accessed 18-01-2020.

15 Berger, L. R. et al. 2015. *Homo naledi*, a New Species of the Genus *Homo* from the Dinaledi Chamber, South Africa. *eLife* 4(e09560).

16 Dirks, P. H. et al. 2017. The Age of *Homo naledi* and Associated Sediments in the Rising Star Cave, South Africa. *eLife* 6(e24231).

17 Durvasula, A. and Sankararaman, S. 2020. Recovering Signals of Ghost Archaic Introgression in African Populations. *Science Advances* 6(7): eaax5097.

18 Harvati, K. et al. 2011. The Later Stone Age Calvaria from Iwo Eleru, Nigeria: Morphology and Chronology. *PLoS ONE*, 6(9): e24024.

19 Grün, R. et al. 2020. Dating the Skull from Broken Hill, Zambia, and Its Position in Human Evolution. *Nature* 580: 372–5. https://doi.org/10.1038/s41586-020-2165-4.

20 Mellars, P. 2006. Going East: New Genetic and Archaeological Perspectives on the Modern Human Colonization of Eurasia. *Science* 313: 796–800.

21 Klein, R. G. 2000, Archeology and the Evolution of Human Behavior. *Evolutionary Anthropology* 9: 17–36.

22 Soares, P. et al. 2012. The Expansion of mtDNA Haplogroup L3 Within and Out of Africa. *Molecular Biology and Evolution* 29(3): 915–27.

23 Groucutt, H. S. et al. 2015. Rethinking the Dispersal of *Homo sapiens* Out of Africa. *Evolutionary Anthropology* 24(4): 149–64.

24 Clarkson, C. et al. 2017. Human Occupation of Northern Australia by 65,000 Years Ago. *Nature* 547: 306–10.

25 Groucutt, H. S. et al. 2015. Stone Tool Assemblages and Models for the Dispersal of *Homo sapiens* Out of Africa. *Quaternary International* 382: 8–30.

26 Liu, W. et al. 2015. The Earliest Unequivocally Modern Humans in Southern China. *Nature* 526: 696–9.

27 Harvati, K. et al. 2019. Apidima Cave Fossils Provide Earliest Evidence of *Homo sapiens* in Eurasia. *Nature* 571: 500–504.

28 Hershkovitz, I. et al. 2018. The Earliest Modern Humans Outside Africa. *Science* 359: 456–9.

29 Petraglia, M. D., Breeze, P. S. and Groucutt, H. S. 2019. Blue Arabia, Green Arabia: Examining Human Colonisation and Dispersal Models. In N. Rasul and I. Stewart (eds.), *Geological Setting, Palaeoenvironment and Archaeology of the Red Sea*, Cham: Springer, pp. 675–83.

30 Drake, N. A. et al. 2011. Ancient Watercourses and Biogeography of the Sahara Explain the Peopling of the Desert. *Proceedings of the National Academy of Sciences* 108(2): 458–62.

31 Groucutt, H. S. et al. 2018. *Homo sapiens* in Arabia by 85,000 Years Ago. *Nature Ecology and Evolution* 2: 800–809.

32 Beyer, R. M. et al. 2020. Windows Out of Africa: A 300,000-Year Chronology of Climatically Plausible Human Contact with Eurasia. *bioRxiv*, preprint online 14 Jan. 2020; http://dx.doi.org/10.1101/2020.01.12.901694.

33 Mallick, S. et al. 2016. The Simons Genome Diversity Project: 300 Genomes from 142 Diverse Populations. *Nature* 538: 201–6.

34 Ibid.; Pagani, L. et al. 2016. Genomic Analyses Inform on Migration Events During the Peopling of Eurasia. *Nature* 538: 238–42.

35 Kuhlwilm, M. et al. 2016. Ancient Gene Flow from Early Modern Humans into Eastern Neanderthals. *Nature* 530: 429–33.

36 Grove, M., Pearce, E. and Dunbar, R. I. 2012. Fission-Fusion and the Evolution of Hominin Social Systems. *Journal of Human Evolution* 62(2): 191–200.

37 His biographer, Judith Heimann, titled her book *The Most Offending Soul Alive: Tom Harrisson and His Remarkable Life*; the title says it all.

38 Members of the Niah team found that pollen from *Justicia* plants occur, in high frequency, coincident with evidence for the presence of forest phases in the jungle. Today, *Justicia* is always the first plant to recolonize following fires in the Niah National Park, so high proportions of its pollen over time might well reflect the use of fire to clear forests.

39 Barker, G. et al. 2007. The 'Human Revolution' in Lowland Tropical Southeast Asia: The Antiquity and Behavior of Anatomically Modern

Humans at Niah Cave (Sarawak, Borneo). *Journal of Human Evolution* 52: 243–61.

40 Wedage, O. et al. 2019. Specialized Rainforest Hunting by *Homo sapiens* ~45,000 Years Ago. *Nature Communications* 10, 739.

41 Langley, M. C. et al. 2020. Bows and Arrows and Complex Symbolic Displays 48,000 Years Ago in the South Asian Tropics. *Science Advances* 6(24): eaba3831.

42 Westaway, K. et al. 2017. An Early Modern Human Presence in Sumatra 73,000–63,000 Years Ago. *Nature* 548: 322–5. Kira Westaway and her colleagues worked at the Lida Ajer site in Sumatra, where previous nineteenth-century researchers had discovered fossil hominin teeth and the bones of extinct animals. Despite physical anthropologists later identifying one particular tooth as that of a modern human, the site was largely ignored because of the lack of a reliable chronology and the fact that it was excavated a long time earlier. Westaway and her team decided to look at it with new scientific approaches. A range of new dating techniques and a new analysis of the tooth showed that it was clearly modern human and that it dated to 63–73,000 years ago.

43 Roberts, P. and Stewart, B. A. 2018. Defining the 'Generalist Specialist' Niche for Pleistocene *Homo sapiens*. *Nature Human Behaviour* 2: 542–50.

44 Bae, C. J., Douka, K. and Petraglia, M. D. 2017. On the Origin of Modern Humans: Asian Perspectives. *Science* 358(6368): eaai9067.

45 Langley et al. 2020.

46 Dennell, R. 2017. Human Colonization of Asia in the Late Pleistocene: The History of an Invasive Species. *Current Anthropology* 58 (Supplement 17): S383–S396.

47 Shipman, P. 2015. *The Invaders: How Humans and Their Dogs Drove Neanderthals to Extinction*, Cambridge, MA: Harvard University Press.

Chapter 3: Neanderthals Emerge into the Light

1 Menez, A. 2018. The Gibraltar Skull: Early History, 1848–1868. *Archives of Natural History* 45.1: 92–110.

2 Schmitz, R. and Thissen, J. 2002. *Neandertal: Die Geschichte geht weiter*, Heidelberg: Spektrum; Madison, P. 2016. The Most Brutal of Human Skulls: Measuring and Knowing the First Neanderthal. *British Journal for the History of Science* 49(3): 411–32.

3 The Feldhofer remains were not the first Neanderthal bones to be found. There were two earlier discoveries: in 1829, at the site of Engis in Belgium, diggers found a child cranium that was later identified as a Neanderthal, and, in Gibraltar, the famous Forbes' Quarry skull was found, it is thought, around 1848, and sent to England for study in 1865 (Menez, 2018).

4 Trinkaus, E. 1985. Pathology and the Posture of the La Chapelle-aux-Saints Neandertal. *American Journal of Physical Anthropology* 67(1): 19–41.

5 Richards, M. P. and Trinkaus, E. 2009. Isotopic Evidence for the Diets of European Neanderthals and Early Modern Humans. *Proceedings of the National Academy of Sciences* 106: 16034–9.

6 Power, R. C. et al. 2018. Dental Calculus Indicates Widespread Plant Use within the Stable Neanderthal Dietary Niche. *Journal of Human Evolution* 119: 27–41.

7 Zilhão, J. et al. 2010. Symbolic Use of Marine Shells and Mineral Pigments by Iberian Neandertals. *Proceedings of the National Academy of Sciences* 107: 1023–8; Stringer, C. et al. 2008. Neanderthal Exploitation of Marine Mammals in Gibraltar. *Proceedings of the National Academy of Sciences* 105: 14319–24.

8 Harvati, K. et al. 2013. New Neanderthal Remains from Mani Peninsula, Southern Greece: The Kalamakia Middle Paleolithic Cave Site. *Journal of Human Evolution* 64: 486–99; Hayden, B. 2012. Neanderthal Social Structure? *Oxford Journal of Archaeology* 31(1): 1–26.

9 Power et al. 2018.

10 Henry, A. G., Brooks, A. and Piperno, D. 2011. Microfossils in Calculus Demonstrate Consumption of Plants and Cooked Foods in Neanderthal Diets (Shanidar III, Iraq; Spy I and II, Belgium). *Proceedings of the National Academy of Sciences* 108(2): 486–91.

11 Rosas, A. et al. 2013. Identification of Neandertal Individuals in Fragmentary Fossil Assemblages by Means of Tooth Associations: The Case of El Sidrón (Asturias, Spain). *Comptes Rendus Palevol* 12(5): 279–91.

12 Defleur, A. et al. 1999. Neanderthal Cannibalism at Moula-Guercy, Ardèche, France. *Science* 286: 128–31.

13 Bar-Yosef, O. 2004. Eat What is There: Hunting and Gathering in the World of Neanderthals and Their Neighbours. *International Journal of Osteoarchaeology* 14: 333–42.

14 Hardy, K. et al. 2012. Neanderthal Medics? Evidence for Food, Cooking, and Medicinal Plants Entrapped in Dental Calculus. *Naturwissenschaften* 99: 617–26.

15 Weyrich, L. et al. 2017. Neanderthal Behaviour, Diet, and Disease Inferred from Ancient DNA in Dental Calculus. *Nature* 544: 357–61.

16 Boëda, É. et al. 2008. New Evidence for Significant Use of Bitumen in Middle Palaeolithic Technical Systems at Umm el Tlel (Syria) Around 70,000 BP. *Paléorient* 34(2): 67–83.

17 Soressi, M. et al. 2013. Neandertals Made the First Specialized Bone Tools in Europe. *Proceedings of the National Academy of Sciences* 110(35): 14186–90.

18 Schoch, W. H. et al. 2015. New Insights on the Wooden Weapons from the Paleolithic Site of Schöningen. *Journal of Human Evolution* 89: 214–25.

19 Migliano, A. B. et al. 2020. Hunter-Gatherers from Different Bands Form Fluid Social Networks That Facilitate Cultural Innovation Through Recombination of Cultural Traditions. *Science Advances* 6(9): eaax5913.

20 Zilhão et al. 2010.

21 Hoffmann, D. L. et al. 2018. Symbolic Use of Marine Shells and Mineral Pigments by Iberian Neandertals 115,000 years ago. *Science Advances* 4(2): eaar5255.

22 Henshilwood, C. S. et al. 2011. A 100,000-Year-Old Ochre-Processing Workshop at Blombos Cave, South Africa. *Science* 334: 219–22.

23 Majkić, A. et al. 2017. A Decorated Raven Bone from the Zaskalnaya VI (Kolosovskaya) Neanderthal Site, Crimea. *PLoS ONE* 12(3): e0173435.

24 Morin, E. and Laroulandie, V. 2012. Presumed Symbolic Use of Diurnal Raptors by Neanderthals. *PLoS ONE* 7(3): e32856.

25 Radovčić, D. et al. 2015. Evidence for Neandertal Jewelry: Modified White-Tailed Eagle Claws at Krapina. *PLoS ONE* 10(3): e0119802.

26 Radovčić, D. et al. 2020. Surface Analysis of an Eagle Talon from Krapina. *Scientific Reports* 10: 6329.

27 Rouzaud, F. 1997. La paléospéléologie ou: l'approche globale des documents anthropiques et paléontologiques conservés dans le karst profond. *Quaternaire* 8(2–3): 257–65. Jaubert, J. 2016. Early Neanderthal Constructions Deep in Bruniquel Cave in Southwestern France. *Nature* 534(7605): 111–14.

28 www.theatlantic.com/science/archive/2016/05/the-astonishing-age-of-a-neanderthal-cave-construction-site/484070/.

29 Rodríguez-Vidal, J. et al. 2014. A Rock Engraving Made by Neanderthals in Gibraltar. *Proceedings of the National Academy of Sciences* 111(37): 13301–6.

30 Hoffmann, D. L. et al. 2018. U-Th Dating of Carbonate Crusts Reveals Neandertal Origin of Iberian Cave Art. *Science* 359: 912–15.

31 D'Errico, F. et al. 1998. Neanderthal Acculturation in Western Europe? A Critical Review of the Evidence and Its Interpretation. *Current Anthropology* 39: S1–S44.

32 Vanhaeren, M. and d'Errico, F. 2006. Aurignacian Ethno-Linguistic Geography of Europe Revealed by Personal Ornaments. *Journal of Archaeological Science* 33: 1105–28.

33 Caron, F. et al. 2011. The Reality of Neandertal Symbolic Behavior at the Grotte du Renne, Arcy-sur-Cure, France. *PLoS ONE* 6(6): e21545.

34 Hublin, J.-J. et al. 2012. Radiocarbon Dates from the Grotte du Renne and Saint-Césaire Support a Neandertal Origin for the Châtelperronian. *Proceedings of the National Academy of Sciences* 109(46): 18743–8.

35 Gravina, B. et al. 2018. No Reliable Evidence for a Neanderthal-Châtelperronian Association at La Roche-à-Pierrot, Saint-Césaire. *Scientific Reports* 8: 15134.

36 Mellars, P. 2005. The Impossible Coincidence. A Single-Species Model for the Origins of Modern Human Behavior in Europe. *Evolutionary Anthropology* 14: 12–27.

37 Smith, F., Falsetti, A. and Donnelly, S. 1989. Modern Human Origins. *Yearbook of Physical Anthropology* 32: 35–68.

38 Krings, M. et al. 1997. Neandertal DNA Sequences and the Origin of Modern Humans. *Cell* 90(1): 19–30; Serre, D. et al. 2004. No Evidence

of Neandertal mtDNA Contribution to Early Modern Humans. *PLoS Biology* 2(3): 313–17.

39 Green, R. E. et al. 2010. A Draft Sequence of the Neandertal Genome. *Science* 328: 710–22.

40 Rozzi, F. V. R. et al. 2009. Cutmarked Human Remains Bearing Neandertal Features and Modern Human Remains Associated with the Aurignacian at Les Rois. *Journal of Anthropological Sciences* 87: 153–85.

Chapter 4: The Road to Denisova Cave

1 The story of the name of Denisova Cave is difficult to disentangle, revolving as it does around oral stories and an absence of reliable local historical records. The literal translation in Russian is 'a cave where Denis lives', but this could refer to Denis or Denisov. The oral tradition relates that: 'The cave got its name in the late eighteenth century when it became a shelter for Dionysius (or Denis), an Old Believer who lived there as a hermit and was a spiritual advisor for the Old Believers from the nearby villages.' 'Old Believers' in this context refers to the split in the Russian Orthodox Church in the mid-seventeenth century into an official church and the Old Believers movement. In the late eighteenth century some settlements of the Old Believers appeared in the Altai region. Another origin story refers to a shepherd named Denis who used the cave to shelter his flock of sheep in bad weather. Michael Shunkov, co-director of the excavations at Denisova, has said that the name might refer to a Denisov family that lived in the area before the October Revolution and worked a water mill on the Anui river. I am grateful to Vladimir Vaneev (SB-RAS) for his work and advice on this. (PS – I favour the monk story.)

2 Slon, V. et al. 2017. A Fourth Denisovan Individual. *Science Advances* 3(7): e1700186.

3 Denisova 1, a small broken tooth, was initially identified as an upper first human incisor, but later analysis revealed it was in fact from a large bovid, perhaps a bison. See, Viola, B. et al. 2011. Late Pleistocene Hominins from the Altai Mountains, Russia. In A. P. Derevianko and M. V. Shunkov (eds.), *Characteristic Features of the Middle to Upper Paleolithic Transition in Eurasia*, Novosibirsk: SB-RAS, pp. 207–13.

4 Weber, A. W. et al. 2006. Radiocarbon Dates from Neolithic and Bronze Age Hunter-Gatherer Cemeteries in the Cis-Baikal Region of Siberia. *Radiocarbon* 48(1): 127–66.

5 Hofreiter, M. et al. 2004. Lack of Phylogeography in European Mammals Before the Last Glaciation. *Proceedings of the National Academy of Sciences* 101(35): 12963–8.

6 Despite the often-fragmented nature of the bones in the site it is possible to identify more than twenty-five species of large mammals and more than 100 small vertebrate species such as birds, fish, amphibians and reptiles. Russian scientists have reconstructed local environmental conditions by exploring the types of small mammals and birds that live today in the various altitudinal levels of the Altai's alpine and sub-alpine communities. Through trapping animals that live today in different ecotones they have explored the climate and environmental preferences of modern fauna. By comparing these data with the archaeological bone and pollen assemblage, they can infer the probable type of environmental conditions that were present in different prehistoric periods.

Chapter 5: The Genetic Revolution

1 Hagelberg, E., Hofreiter, M. and Keyser, C. 2015. Ancient DNA: The First Three Decades. *Philosophical Transactions of the Royal Society B* 370: 2013.0371.

2 Cooper, A. and Poinar, H. N. 2000. Ancient DNA: Do It Right or Not at All. *Science* 289: 1139.

3 Madison, P. 2016. The Most Brutal of Human Skulls: Measuring and Knowing the First Neanderthal. *British Journal for the History of Science* 49(3): 411–32.

4 Cooper, A. et al. 2001. Human Origins and Ancient Human DNA. *Science* 292: 1655–6.

5 Dabney, J., Meyer, M. and Pääbo, S. 2013. Ancient DNA Damage. *Cold Spring Harbor Perspectives in Biology* 5(7): a012567. https://doi.org/10.1101/cshperspect.a012567.

6 Gansauge, M.-T. and Meyer, M. 2014. Selective Enrichment of Damaged DNA Molecules for Ancient Genome Sequencing. *Genome Research* 24(9): 1543–9.

7 Korlević, P. et al. 2015. Reducing Microbial and Human Contamination in DNA Extractions from Ancient Bones and Teeth. *BioTechniques* 59(2): 87–93.

8 Mullis was an interesting and complex character. He claimed that LSD had a major role in his work life, and in his autobiography, *Dancing Naked in the Mind Field*, he wrote of an occasion near his cabin in the woods in which he encountered a glowing green raccoon, potentially an alien, that said 'Good evening, Doctor' to him. He came up with the idea for PCR in a classic Eureka moment while driving late at night from Berkeley to his cabin in the woods on California's Highway 128. He pulled off the road at mile-marker 46.58 and, while his girlfriend slept next to him, scribbled down notes as he thought up the method. He realized that PCR could be massively important, enabling the amplification of huge amounts of DNA. He was convinced that night that he would one day win the Nobel Prize for this development. Sadly, his later career was marked by the denial of climate change and the false idea that HIV did not cause AIDS.

9 Green, R. E. et al. 2010. A Draft Sequence of the Neandertal Genome. *Science* 328: 710–22.

10 For a general description, see www.yourgenome.org/facts/what-is-the-454-method-of-dna-sequencing.

11 Rasmussen, M. et al. 2010. Ancient Human Genome Sequence of an Extinct Palaeo-Eskimo. *Nature* 463: 757–62.

12 Olalde, I. et al. 2018. The Beaker Phenomenon and the Genomic Transformation of Northwest Europe. *Nature* 555: 190–96.

Chapter 6: A New Species of Human

1 Pääbo, S. 2014. *Neanderthal Man: In Search of Lost Genomes*. New York: Basic Books.

2 Ibid.

3 Krause, J. et al. 2010. The Complete Mitochondrial DNA Genome of an Unknown Hominin from Southern Siberia. *Nature*. 464: 894–7.

4 Reich, D. et al. 2010. Genetic History of an Archaic Hominin Group from Denisova Cave in Siberia. *Nature* 468: 1053–60.

5 Meyer, M. et al. 2012. A High-Coverage Genome Sequence from an Archaic Denisovan Individual. *Science* 338: 222–6. A high-coverage genome is generally >20X coverage. The record holder for ancient genomes is held by a Swedish Mesolithic genome at 70X coverage from the site of Stora Förvar (published by Torsten Günther and colleagues in *PLoS Biology* 2018). Amongst our older Palaeolithic specimens, the Denisova 5 'Altai Neanderthal' bone stands out at an incredible 52X genome. This is a higher coverage than the majority of genomes from the highest covered living humans. The genome of Craig Venter, for example, head of one of the two teams that raced to be the first to sequence the human genome, has coverage of 7.5X.

6 Green, R. E. et al. 2010. A Draft Sequence of the Neandertal Genome. *Science* 328: 710–22.

7 Bennett, E. A. et al. 2019. Morphology of the Denisovan Phalanx Closer to Modern Humans Than to Neanderthals. *Science Advances* 4 September 2019: eaaw3950.

8 Meyer et al. 2012.

9 Prüfer, K. et al. 2017. A High-Coverage Neandertal Genome from Vindija Cave in Croatia. *Science* 358: 655–8.

10 Green et al. 2010.

11 Reich, D. et al. 2010.

Chapter 7: Where are the Fossil Remains?

1 Sawyer, S. et al. 2015. Nuclear and Mitochondrial DNA Sequences from Two Denisovan Individuals. *Proceedings of the National Academy of Sciences* 112: 15696–700.

2 Prüfer, K. et al. 2017. A High-Coverage Neandertal Genome from Vindija Cave in Croatia. *Science* 358: 655–8.

3 Bennett, E. A. et al. 2019. Morphology of the Denisovan Phalanx Closer to Modern Humans Than to Neanderthals. *Science Advances* 4 September 2019: eaaw3950.

4 Reich, D. et al. 2010. Genetic History of an Archaic Hominin Group from Denisova Cave in Siberia. *Nature* 468: 1053–60.

5 See Meyer, M. et al. 2012. A High-Coverage Genome Sequence from an Archaic Denisovan Individual. *Science* 338: 222–6. The section from which this analysis comes is by Viola, B. et al. Morphology of the Denisova Molar and Phalanx. Stratigraphy and Dating (Supplementary Information 12).

6 Arnold, L. J. et al. 2014. Luminescence Dating and Palaeomagnetic Age Constraint on Hominins from Sima de los Huesos, Atapuerca, Spain. *Journal of Human Evolution* 67: 85–107.

7 Xing, S. et al. 2015. Hominin Teeth from the Early Late Pleistocene Site of Xujiayao, Northern China. *American Journal of Physical Anthropology* 156: 224–40; Martinón-Torres, M. et al. 2018. A 'Source and Sink' Model for East Asia? Preliminary Approach Through the Dental Evidence. *Comptes Rendus Palevol* 17: 1–2, 33–43.

8 Tu, H. et al. 2015. ^{26}Al/^{10}Be Burial Dating of Xujiayao-Houjiayao Site in Nihewan Basin, Northern China. *PLoS ONE* 10(2): e0118315.

9 Xing et al. 2015.

10 Martinón-Torres, M. et al. 2017. *Homo sapiens* in the Eastern Asian Late Pleistocene. *Current Anthropology* 58(S17): S434–S448.

11 Li Zhan-Yang of the Institute of Vertebrate Paleontology and Paleoanthropology (IVPP) in Beijing was the excavator.

12 Li, Z. et al. 2019. Engraved Bones from the Archaic Hominin Site of Lingjing, Henan Province. *Antiquity* 93(370): 886–900.

13 Li, Z.-Y. et al. 2017. Late Pleistocene Archaic Human Crania from Xuchang, China. *Science* 355: 969–72.

14 Llamas, B. et al. 2012. High-Resolution Analysis of Cytosine Methylation in Ancient DNA. *PLoS ONE* 7: e30226.

15 Gokhman, D. et al. 2014. Reconstructing the DNA Methylation Maps of the Neandertal and the Denisovan. *Science* 344: 523–7.

16 Briggs, A. W. et al. 2007. Patterns of Damage in Genomic DNA Sequences from a Neandertal. *Proceedings of the National Academy of Sciences* 104(37): 14616–21.

17 Gokhman, D. et al. 2019. Reconstructing Denisovan Anatomy Using DNA Methylation Maps. *Cell* 179(1): 180–92.

18 Gokhman et al. 2019.

19 Schneider, E., El Hajj, N. and Haaf, T. 2014. Epigenetic Information from Ancient DNA Provides New Insights into Human Evolution. Commentary on Gokhman, D. et al. (2014): Reconstructing the DNA Methylation Maps of the Neanderthal and the Denisovan. *Science* 344: 523–7. *Brain, Behavior and Evolution* 84(3): 169–71.

20 Dongju Zhang, pers. comm., 22 November 2019.

21 www.eva.mpg.de/evolution/downloads.html.

22 Welker, F. et al. 2016. Palaeoproteomic Evidence Identifies Archaic Hominins Associated with the Châtelperronian at the Grotte du Renne. *Proceedings of the National Academy of Sciences* 113(40): 11162–7; Welker, F. 2018. Palaeoproteomics for Human Evolution Studies. *Quaternary Science Reviews* 190: 137–47.

23 Orlando, L. et al. 2013. Recalibrating *Equus* Evolution Using the Genome Sequence of an Early Middle Pleistocene Horse. *Nature* 499: 74–8.

24 Demarchi, B. et al. 2016. Protein Sequences Bound to Mineral Surfaces Persist into Deep Time. *eLife* 5: e17092.

25 Schroeter, E. R. et al. 2017. Expansion for the *Brachylophosaurus canadensis* Collagen I Sequence and Additional Evidence of the Preservation of Cretaceous Protein. *Journal of Proteome Research* 16(2): 920–32.

26 Welker, F. et al. 2020. The Dental Proteome of *Homo antecessor*. *Nature* 580: 235–8; Welker, F. et al. 2019. Enamel Proteome Shows That *Gigantopithecus* Was an Early Diverging Pongine. *Nature* 576(7786): 262–5.

27 Chen, F. et al. 2019. A Late Middle Pleistocene Denisovan Mandible from the Tibetan Plateau. *Nature* 569: 409–12.

28 Bailey, S., Hublin, J.-J. and Antón, S. 2019. Rare Dental Trait Provides Morphological Evidence of Archaic Introgression in Asian Fossil Record. *Proceedings of the National Academy of Sciences* 116(30): 14806–7.

29 Ibid.

30 Interestingly, in populations like the Nepalese, the presence of three-rooted molars (3RM) is high, at around 25 per cent, consistent with the presence of other Denisovan-derived traits such as the *EPAS1* gene. In other parts of the Denisovan world, however, such as Papua New

Guinea and Australia, the 3RM incidence is lower: 12 per cent in Australian Aborigines. Some have suggested that there might be an effect on the presence of this trait through influences such as masticatory robustness (see Bailey, Hublin and Antón 2019).

31 Chang, C.-H. et al. 2015. The First Archaic *Homo* from Taiwan. *Nature Communications* 6, Article 6037.

32 Viola, B. et al. 2019. A Parietal Fragment from Denisova Cave. Abstract: 88th Annual Meeting of the American Association of Physical Anthropologists, Cleveland, OH.

Chapter 8: Finding Needles in Haystacks

1 Buckley, M. et al. 2009. Species Identification by Analysis of Bone Collagen Using Matrix-Assisted Laser Desorption/Ionisation Time-of-Flight Mass Spectrometry. *Rapid Communications in Mass Spectrometry* 23: 3843–54.

2 Welker, F. et al. 2015. Using ZooMS to Identify Fragmentary Bone from the Late Middle/Early Upper Palaeolithic Sequence of Les Cottés, France. *Journal of Archaeological Science* 54: 279–86.

3 Brown, S. et al. 2016. Identification of a New Hominin Bone from Denisova Cave, Siberia Using Collagen Fingerprinting and Mitochondrial DNA Analysis. *Scientific Reports* 6, Article 23559. https://doi.org/10.1038/srep23559.

4 Slon, V. et al. 2018. The Genome of the Offspring of a Neanderthal Mother and a Denisovan Father. *Nature* 561(7721): 113–16.

5 www.nature.com/articles/d41586-018-06004-0.

6 Prüfer, K. et al. 2014. The Complete Genome Sequence of a Neanderthal from the Altai Mountains. *Nature* 505(7481): 43–9.

7 We have continued to use ZooMS at Denisova Cave through Katerina's ongoing work and Sam's PhD. As of June 2020 we have identified a total of nine new hominin bones from about 10,000 bone fragments; roughly one in a thousand. In the space of four years ZooMS has more than doubled the total number of human remains we have from Denisova. Denisova had an estimated 135,000 bones, of which 128,000 were unidentified between 1986 and 2010, so we have the chance of finding

perhaps 100–110 bone fragments of hominins from Denisova, a site occupied for around 200,000 years.

Chapter 9: The Science of 'When'

1 Libby, W. F., Anderson, E. C. and Arnold, J. R. 1949. Age Determination by Radiocarbon Content: World-Wide Assay of Natural Radiocarbon. *Science* 109: 227–8.
2 Higham, T. F. G. et al. 2014. The Timing and Spatiotemporal Patterning of Neanderthal Disappearance. *Nature* 512: 306–9.
3 Reich, D. et al. 2010. Genetic History of an Archaic Hominin Group from Denisova Cave in Siberia. *Nature* 468: 1053–60.
4 Michael Balter wrote a nice review of their work in *Science* magazine in 2011: https://science.sciencemag.org/content/332/6030/658.1.summary.
5 McGrayne, S. B. 2011. *The Theory That Would Not Die: How Bayes' Rule Cracked the Enigma Code, Hunted Down Russian Submarines, & Emerged Triumphant from Two Centuries of Controversy.* New Haven: Yale University Press.
6 Buck, C. E., Cavanagh, W. G. and Litton, C. D. 1996. *Bayesian Approach to Interpreting Archaeological Data.* Chichester: John Wiley and Sons.
7 Douka, K. et al. 2019. Age Estimates for Hominin Fossils and the Onset of the Upper Palaeolithic at Denisova Cave. *Nature* 565: 640–44.

Chapter 10: On the Trail of the Modern Human Diaspora

1 The distribution shown in Figure 23 is based on Kuhn, S. and Zwyns, N. 2014. Rethinking the Initial Upper Paleolithic. *Quaternary International* 347: 29–38.
2 Belfer-Cohen, A. and Goring-Morris, A. N. 2017. The Upper Palaeolithic in Cisjordan. In Y. Enzel and O. Bar-Yosef (eds.), *Quaternary of the Levant: Environments, Climate Change, and Humans,* Cambridge: Cambridge University Press, pp. 627–37.
3 Rose, J. I. and Marks, A. E. 2014. 'Out of Arabia' and the Middle-Upper Palaeolithic Transition in the Southern Levant. *Quartär* 61: 49–85.

4 Bergman, C. A. and Stringer, C. B. 1989. Fifty Years After: Egbert, an Early Upper Palaeolithic Juvenile from Ksar Akil, Lebanon. *Paléorient* 15(2): 99–111.

5 This is based on a letter from Fr Franklin Ewing to Dr John Otis Brew, cc'd to Professor Hallam Movius, outlining an account of his remembrance of what happened in Lebanon, June 1966. I am grateful to Dr Katerina Douka for locating this in the archives of the Peabody Museum of Archaeology and Ethnology, Cambridge MA.

6 Bolton, B. 1953. The Great Adventures of Egbert! *The Ram* (Fordham College newspaper) 33–5 (29 October 1953): 3–5. This document shows that Egbert was returned to the Lebanon by ship on 28 October 1953.

7 Bergman and Stringer 1989.

8 Kuhn and Zwyns 2014.

9 Dalén, L. et al. 2012. Partial Genetic Turnover in Neandertals: Continuity in the East and Population Replacement in the West. *Molecular Biology and Evolution* 29(8): 1893–7.

10 Müller, U. C. et al. 2011. The Role of Climate in the Spread of Modern Humans into Europe. *Quaternary Science Reviews* 30(3–4): 273–9.

11 Shunkov, M. V., Kozlikin, M. B. and Derevianko, A. P. 2020. Dynamics of the Altai Paleolithic Industries in the Archaeological Record of Denisova Cave. *Quaternary International* https://doi.org/10.1016/j.quaint.2020.02.017.

12 Moorjani, P. et al. 2016. A Genetic Method for Dating Ancient Genomes Provides a Direct Estimate of Human Generation Interval in the Last 45,000 Years. *Proceedings of the National Academy of Sciences* 113(20): 5652–7; Racimo, F., et al. 2015. Evidence for Archaic Adaptive Introgression in Humans. *Nature Reviews Genetics* 16: 359–71 (Figure 28 is based on this article).

13 Fu, Q. et al. 2014. The Genome Sequence of a 45,000-Year-Old Modern Human from Western Siberia. *Nature* 514: 445–9.

Chapter 11: DNA from Dirt

1 Willerslev, E. et al. 2007. Ancient Biomolecules from Deep Ice Cores Reveal a Forested Southern Greenland. *Science* 317: 111–14.

2 Willerslev, E. et al. 2003. Diverse Plant and Animal Genetic Records from Holocene and Pleistocene Sediments. *Science* 300: 791–5.

3 Hofreiter, M. et al. 2003. Molecular Caving. *Current Biology* 13(18): R693–R695.

4 Haile, J. et al. 2007. Ancient DNA Chronology Within Sediment Deposits: Are Paleobiological Reconstructions Possible and is DNA Leaching a Factor? *Molecular Biology and Evolution* 24: 982–9.

5 Slon, V. et al. 2017. Neandertal and Denisovan DNA from Pleistocene Sediments. *Science* 356: 605–8.

6 Ibid.

7 Münzel, S., Seeberger, F. and Hein, W. 2002. The Geißenklösterle Flute: Discovery, Experiments, Reconstruction. *Studien zur Musikarchäologie* III: 107–18.

8 Conard, N., Malina, M. and Münzel, S. 2009. New Flutes Document the Earliest Musical Tradition in Southwestern Germany. *Nature* 460: 737–40.

9 Zavala, E. et al. 2019. Recovery of Ancient Hominin and Mammalian Mitochondrial DNA from High Resolution Screening of Pleistocene Sediments at Denisova Cave. ESHE Meeting Abstract book, Liège, Belgium.

Chapter 12: *The Hobbits*

1 Lambeck, K., Yokoyama, Y. and Purcell, T. 2002. Into and Out of the Last Glacial Maximum: Sea-Level Change During Oxygen Isotope Stages 3 and 2. *Quaternary Science Reviews* 21: 343–60.

2 Cooper, A. and Stringer, C. B. 2013. Did the Denisovans Cross Wallace's Line? *Science* 342(6156): 321–3.

3 Foster, J. B. 1964. Evolution of Mammals on Islands. *Nature* 202: 234–5.

4 Morwood, M. J. et al. 1997. Stone Artefacts from the 1994 Excavation at Mata Menge, West Central Flores, Indonesia. *Australian Archaeology* 44(1): 26–34.

5 Morwood, M. J. and van Oosterzee, P. 2016. *A New Human: The Startling Discovery and Strange Story of the 'Hobbits' of Flores, Indonesia*. Abingdon: Routledge.

6 Brown, P. et al. 2004. A New Small-Bodied Hominin from the Late Pleistocene of Flores, Indonesia. *Nature* 431(7012): 1055–61.

7 Morwood and van Oosterzee 2016.

8 Callaway, E. 2014. The Discovery of *Homo floresiensis*: Tales of the Hobbit. *Nature* 514: 422–6. The story about '*floresianus*' is from Henry Gee.

9 Brown, P. and Morwood, M. J. 2004. Some Initial Informal Reactions to Publication of the Discovery of *Homo floresiensis* and Replies from Brown and Morwood. *Before Farming* 4: 1–7.

10 *Guardian*, 13 January 2005; www.theguardian.com/science/2005/jan/13/research.science.

11 Morwood, M. J. et al. 2005. Further Evidence for Small-Bodied Hominins from the Late Pleistocene of Flores, Indonesia. *Nature* 437: 1012–17.

12 Tocheri, M. W. et al. 2007. The Primitive Wrist of *Homo floresiensis* and Its Implications for Hominin Evolution. *Science* 317: 1743–5. For a popular account of some of this work see Tocheri, M. 2007. Joining in Fellowship with the Hobbits. *AnthroNotes: Museum of Natural History Publication for Educators* 28(2): 1–5.

13 Sutikna, T. et al. 2016. Revised Stratigraphy and Chronology for *Homo floresiensis* at Liang Bua in Indonesia. *Nature* 532: 366–9.

14 Kaifu, Y. et al. 2015. Unique Dental Morphology of *Homo floresiensis* and Its Evolutionary Implications. *PLoS ONE* https://doi.org/10.1371/journal.pone.0141614.

15 Van den Bergh, G. D. et al. 2016. *Homo floresiensis*-like Fossils from the Early Middle Pleistocene of Flores. *Nature* 534: 245–8.

16 Brumm, A. et al. 2010. Hominins on Flores, Indonesia, by One Million Years Ago. *Nature* 464: 748–52.

17 Kubo, D., Kono, R. T. and Kaifu, Y. 2013. Brain Size of *Homo floresiensis* and Its Evolutionary Implications. *Proceedings of the Royal Society B* 280: 2013.0338.

18 Argue, D. et al. 2017. The Affinities of *Homo floresiensis* Based on Phylogenetic Analyses of Cranial, Dental and Postcranial Characters. *Journal of Human Evolution* 107: 107–33.

19 Gómez-Robles, A. 2016. The Dawn of *Homo floresiensis*. *Nature* 534: 188–9.

20 Roberts, R. and Sutikna, T. 2013. Michael John Morwood (1950–2013). *Nature* 500: 401. https://doi.org/10.1038/500401a.

21 Mijares, A. S. et al. 2010. New Evidence for a 67,000-Year-Old Human Presence at Callao Cave, Luzon, Philippines. *Journal of Human Evolution* 59: 123–32.

22 Détroit, F. et al. 2019. A New Species of *Homo* from the Late Pleistocene of the Philippines. *Nature* 568: 181–6.

23 Tocheri, M. W. 2019. Previously Unknown Human Species Found in Asia Raises Questions About Early Hominin Dispersals from Africa. *Nature* 568: 176–8.

24 Ibid.

25 Ingicco, T. et al. Earliest Known Hominin Activity in the Philippines by 709 Thousand Years Ago. *Nature* 557: 233–7.

26 There are some strong dissenting voices, e.g. Bednarik, R. G. 2003. Seafaring in the Pleistocene. *Cambridge Archaeological Journal* 13(1): 41–66, and comments therein.

27 Dennell, R. 2020. *From Arabia to the Pacific: How Our Species Colonised Asia*. Oxford: Routledge.

28 Van den Bergh, G. et al. 2016. Earliest Hominin Occupation of Sulawesi, Indonesia. *Nature* 529: 208–11.

Chapter 13: *The Journey to the East of Wallace's Line*

1 Reich, D. et al. 2010. Genetic History of an Archaic Hominin Group from Denisova Cave in Siberia. *Nature* 468: 1053–60.

2 Cooper, A. and Stringer, C. B. 2013. Did the Denisovans Cross Wallace's Line? *Science*, 342(6156): 321–3.

3 Reich, D. et al. 2011. Denisova Admixture and the First Modern Human Dispersals into Southeast Asia and Oceania. *American Journal of Human Genetics* 89: 516–28.

4 Cooper and Stringer 2013.

5 Qin, P. and Stoneking, M. 2015. Denisovan Ancestry in East Eurasian and Native American Populations. *Molecular Biology and Evolution* 32(10): 2665–74.

6 Ibid.

7 Trinkaus, E. and Shang, H. 2008. Anatomical Evidence for the Antiquity of Human Footwear: Tianyuan and Sunghir. *Journal of Archaeological Science* 35(7): 1928–33.

8 Browning, S. R. et al. 2018. Analysis of Human Sequence Data Reveals Two Pulses of Archaic Denisovan Admixture. *Cell* 173(1): 53–61.

9 Prüfer, K. et al. 2017. A High-Coverage Neandertal Genome from Vindija Cave in Croatia. *Science* 358: 655–8.

10 Moore, C. 2003. *New Guinea: Crossing Boundaries and History*. Honolulu: University of Hawai'i Press.

11 Palmer, B. (ed.) 2018. *The Languages and Linguistics of the New Guinea Area: A Comprehensive Guide*. The World of Linguistics 4. Berlin: De Gruyter Mouton.

12 Bergström, A. et al. 2017. A Neolithic Expansion, But Strong Genetic Structure, in the Independent History of New Guinea. *Science* 357: 1160–63.

13 Jacobs, G. et al. 2019. Multiple Deeply Divergent Denisovan Ancestries in Papuans. *Cell* 177(4): 1010–21. https://doi.org/10.1016/j.cell.2019.02.035. The team is led by Murray Cox and they work alongside the Indonesian Genome Diversity Project (IGDP) team based at the Eijkman Institute in Jakarta, aiming to improve knowledge of genetic data from Indonesians.

14 Massilani, D. et al. 2020. Denisovan Ancestry and Population History of Early East Asians. *bioRxiv* 2020.06.03.131995. https://doi.org/10.1101/2020.06.03.131995.

15 Yang, M. A. et al. 2017. 40,000-Year-Old Individual from Asia Provides Insight into Early Population Structure in Eurasia. *Current Biology* 27(20): 3202–8.

16 Jacobs et al. 2019.

17 Malaspinas, A.-S. et al. 2016. A Genomic History of Aboriginal Australia. *Nature* 538: 207–14.

18 GenomeAsia100K Consortium 2019. The GenomeAsia 100K Project Enables Genetic Discoveries Across Asia. *Nature* 576(7785): 106–11.

19 Bouckaert, R. R., Bowern, C. and Atkinson, Q. D. 2018. The Origin and Expansion of Pama–Nyungan Languages Across Australia. *Nature Ecology and Evolution* 2: 741–9.

20 Clarkson, C. et al. 2017. Human Occupation of Northern Australia by 65,000 Years Ago. *Nature* 547: 306–10.

21 O'Connell, J. F. and Allen, J. 2015. The Process, Biotic Impact, and Global Implications of the Human Colonization of Sahul About 47,000 Years Ago. *Journal of Archaeological Science* 56: 73–84; Wood, R. 2017. Comments on the Chronology of Madjedbebe. *Australian Archaeology* 83(3): 172–4.

22 Bird, M. I. et al. 2019. Early Human Settlement of Sahul was Not an Accident. *Scientific Reports* 9: 8220.

23 Thorne, A. G. and Macumber, P. G. 1972. Discoveries of Late Pleistocene Man at Kow Swamp, Australia. *Nature* 238: 316–19.

24 Flood, J. 1995. *Archaeology of the Dreamtime: The Story of Prehistoric Australia and Its People* (revised edn). London: HarperCollins.

25 Stone, T. and Cupper, M. L. 2003. Last Glacial Maximum Ages for Robust Humans at Kow Swamp, Southern Australia. *Journal of Human Evolution* 45: 99–111.

26 Bowler, J. M. et al. 1970. Pleistocene Human Remains from Australia: A Living Site and Human Cremation from Lake Mungo, Western New South Wales. *World Archaeology* 2(1): 39–60.

27 Olley, J. M. et al. 2006. Single-Grain Optical Dating of Grave-Infill Associated with Human Burials at Lake Mungo, Australia. *Quaternary Science Reviews* 25(19–20): 2469–74.

28 Westaway, M. C. and Groves, C. P. 2009. The Mark of Ancient Java is on None of Them. *Archaeology in Oceania* 44: 84–95.

29 Ibid.

30 Hiscock, P. 2008. *Archaeology of Ancient Australia*. Oxford: Routledge.

31 Adcock, G. J. et al. 2001. Mitochondrial DNA Sequences in Ancient Australians: Implications for Modern Human Origins. *Proceedings of the National Academy of Sciences* 98(2): 537–42.

32 Heupink, T. H. et al. 2016. Ancient DNA Sequences from the First Australians Revisited. *Proceedings of the National Academy of Sciences* 113(25): 6892–7.

33 Mulvaney, D. J. 1991. Past Regained, Future Lost: The Kow Swamp Pleistocene Burials. *Antiquity* 65(246): 12–21; Bowdler, S. 1992. Unquiet Slumbers: The Return of the Kow Swamp Burials. *Antiquity* 66(250): 103–6.

34 Gelder, K. and Jacobs, J. M. 1994. *Uncanny Australia: Sacredness and Identity in a Postcolonial Nation*. Melbourne: Melbourne University Publishing.

35 *Sydney Morning Herald*, 18 December 2018.

36 Jacobs et al. 2019; Malaspinas et al. 2016.

37 O'Connor, S. et al. 2017. Hominin Dispersal and Settlement East of Huxley's Line: The Role of Sea Level Changes, Island Size, and Subsistence Behavior. *Current Anthropology* 58(S17): S567–S582.

38 Bradshaw, C. J. A. et al. 2019. Minimum Founding Populations for the First Peopling of Sahul. *Nature Ecology and Evolution* 3: 1057–63.

39 Ibid.

40 Clarkson et al. 2017.

41 O'Connell, J. F. et al. 2018. When Did *Homo sapiens* First Reach Southeast Asia and Sahul? *Proceedings of the National Academy of Sciences* 115(34): 8482–90.

42 Dortch, J. and Malaspinas, A.-S. 2017. Madjedbebe and Genomic Histories of Aboriginal Australia. *Australian Archaeology* 83(3): 174–7.

43 Cooper and Stringer 2013.

Chapter 14: Homo erectus *and the Ghost Population*

1 Prüfer, K. et al. 2017. A High-Coverage Neandertal Genome from Vindija Cave in Croatia. *Science* 358: 655–8.

2 Trinkaus, E. and Shipman, P. 1993. *The Neandertals: Changing the Image of Mankind*. London: Pimlico.

3 Shipman, P. 2001. *The Man Who Found the Missing Link: The Extraordinary Life of Eugène Dubois*. London: Weidenfeld and Nicolson.

4 Zaim, Y. et al. 2011. New 1.5 Million-Year-Old *Homo erectus* Maxilla from Sangiran (Central Java, Indonesia). *Journal of Human Evolution* 61: 363–76.

5 Gathogo, P. N. and Brown, F. H. 2006. Revised Stratigraphy of Area 123, Koobi Fora, Kenya, and New Age Estimates of Its Fossil Mammals, Including Hominins. *Journal of Human Evolution* 51(5): 471–9.

6 Dennell, R. and Roebroeks, W. 2006. An Asian Perspective on Early Human Dispersal from Africa. *Nature* 438: 1099–1104.

7 Tobias, P. V. 1976. The Life and Times of Ralph von Koenigswald: Palaeontologist Extraordinary. *Journal of Human Evolution* 5(5): 403–12.

8 Franzen, J. 1984. G. H. R. von Koenigswald and Asia – An Obituary. *Asian Perspectives* 25(2): 43–51.

9 Matsu'ura, S. et al. 2020. Age Control of the First Appearance Datum for Javanese *Homo erectus* in the Sangiran Area. *Science* 367: 210–14.

10 Von Koenigswald, G. H. R. and Weidenreich, F. 1938. Discovery of an Additional *Pithecanthropus* Skull. *Nature* 142: 715.

11 Thorne, A. G. and Wolpoff, M. H. 1981. Regional Continuity in Australasian Pleistocene Hominid Evolution. *American Journal of Physical Anthropology* 55: 337–49.

12 Their work in Java is told first-hand in Swisher III, C. C., Curtis, G. H. and Lewin, R. 2000. *Java Man: How Two Geologists Changed Our Understanding of Human Evolution.* Chicago: Chicago University Press.

13 Swisher III, C. C. et al. 1996. Latest *Homo erectus* of Java: Potential Contemporaneity with *Homo sapiens* in Southeast Asia. *Science* 274: 1870–74.

14 Rizal, Y. et al. 2020. Last Appearance of *Homo erectus* at Ngandong, Java, 117,000–108,000 Years Ago. *Nature* 577: 381–5.

15 Ibid.

16 Storm, P. et al. 2005. Late Pleistocene *Homo sapiens* in a Tropical Rainforest Fauna in East Java. *Journal of Human Evolution* 49(4): 536–45.

17 Westaway, K. E. et al. 2007. Age and Biostratigraphic Significance of the Punung Rainforest Fauna, East Java, Indonesia, and Implications for *Pongo* and *Homo. Journal of Human Evolution* 53(6): 709–17.

18 See Chapter 2, note 42.

19 Kaifu, Y., Baba, H. and Aziz, F. 2006. Indonesian *Homo erectus* and Modern Human Origins in Australasia: New Evidence from the Sambungmacan Region, Central Java. In Y. Tomida et al. (eds.), *Proceedings of the 7th and 8th Symposia on Collection Building and Natural History Studies in Asia and the Pacific Rim*, National Science Museum Monographs (34), Tokyo: National Science Museum, pp. 289–94.

20 Delson, E. et al. 2001. The Sambungmacan 3 *Homo erectus* Calvaria: A Comparative Morphometric and Morphological Analysis. *Anatomical Record* 262: 380–97.

21 Turney, C. S. M. et al. 2020. Late Survival of Multiple Hominin Species in Island Southeast Asia. Manuscript in submission.

22 Browning, S. R. et al. 2018. Analysis of Human Sequence Data Reveals Two Pulses of Archaic Denisovan Admixture. *Cell* 173(1): 53–61.

23 Lawson, D. J., van Dorp, L. and Falush, D. 2018. A Tutorial on How Not to Over-Interpret STRUCTURE and ADMIXTURE Bar Plots. *Nature Communications* 9(3258).

Chapter 15: Disappearing from the World

1 Sagan, C. 2006. *The Varieties of Scientific Experience: A Personal View of the Search for God*, ed. A. Druyan. Harmondsworth: Penguin.

2 Sutikna, T. et al. 2018. The Spatio-Temporal Distribution of Archaeological and Faunal Finds at Liang Bua (Flores, Indonesia) in Light of the Revised Chronology for *Homo floresiensis*. *Journal of Human Evolution* 124: 52–72.

3 Tucci, S. et al. 2018. Evolutionary History and Adaptation of a Human Pygmy Population of Flores Island, Indonesia. *Science* 361: 511–16.

4 Gravina, B., Mellars, P. and Ramsey, C. 2005. Radiocarbon Dating of Interstratified Neanderthal and Early Modern Human Occupations at the Châtelperronian Type-Site. *Nature* 438: 51–6; Zilhão, J. et al. 2008. Grotte des Fées (Châtelperron): History of Research, Stratigraphy, Dating, and Archaeology of the Châtelperronian Type-Site. *PalaeoAnthropology* 2008: 1–42; Riel-Salvatore, J., Miller, A. E. and Clark, G. A. 2008. An Empirical Evaluation of the Case for a Châtelperronian-Aurignacian Interstratification at Grotte des Fées de Châtelperron. *World Archaeology* 40(4): 480–92.

5 Finlayson, C. et al. 2008. Gorham's Cave, Gibraltar – The Persistence of a Neanderthal Population. *Quaternary International* 181(1): 64–71.

6 Finlayson, C. et al. 2006. Late Survival of Neanderthals at the Southernmost Extreme of Europe. *Nature* 443: 850–53.

7 Banks, W. E. et al. 2008. Neanderthal Extinction by Competitive Exclusion. *PLoS ONE* 3(12): e3972.

8 Wood, R. E. et al. 2013. Radiocarbon Dating Casts Doubt on the Late Chronology of the Middle to Upper Palaeolithic Transition in Southern Iberia. *Proceedings of the National Academy of Sciences* 110(8): 2781–6.

9 Pinhasi, R. et al. 2011. Revised Age of Late Neanderthal Occupation and the End of the Middle Paleolithic in the Northern Caucasus. *Proceedings of the National Academy of Sciences* 108(21): 8611–16.

10 Higham, T. F. G. 2011. European Middle and Upper Palaeolithic Radiocarbon Dates are Often Older Than They Look: Problems with Previous Dates and Some Remedies. *Antiquity* 85(327): 235–49.

11 Higham, T. F. G. et al. 2014. The Timing and Spatiotemporal Patterning of Neanderthal Disappearance. *Nature* 512: 306–9.

12 Trinkaus, E., Constantin, S. and Zilhão, J. (eds.). 2013. *Life and Death at the Peștera cu Oase: A Setting for Modern Human Emergence in Europe.* Oxford: Oxford University Press.

13 Trinkaus, E. et al. 2003. An Early Modern Human from the Peștera cu Oase, Romania. *Proceedings of the National Academy of Sciences* 100(20): 11231–6.

14 Fu, Q. et al. 2015. An Early Modern Human from Romania with a Recent Neanderthal Ancestor. *Nature* 524(7564): 216–19.

15 Benazzi, S. et al. 2011. Early Dispersal of Modern Humans in Europe and Implications for Neanderthal Behaviour. *Nature* 479: 525–8.

16 Hublin, J.-J. et al. 2020. Initial Upper Palaeolithic *Homo sapiens* from Bacho Kiro Cave, Bulgaria. *Nature* 581: 299–302.

17 Hajdinjak, M. et al. 2018. Reconstructing the Genetic History of Late Neanderthals. *Nature* 555: 652–6.

18 Mellars, P. 2004. Neanderthals and the Modern Human Colonization of Europe. *Nature* 432: 461–5.

19 Giaccio, B. et al. 2017. High-Precision ^{14}C and ^{40}Ar/^{39}Ar Dating of the Campanian Ignimbrite (Y-5) Reconciles the Time-Scales of Climatic-Cultural Processes at 40 ka. *Scientific Reports* 7(45940). https://doi.org/10.1038/srep45940.

20 Sinitsyn, A. A. 2003. A Palaeolithic 'Pompeii' at Kostenki, Russia. *Antiquity* 77(295): 9–14.

21 Golovanova, L. V. et al. 2010. Significance of Ecological Factors in the Middle to Upper Paleolithic Transition. *Current Anthropology* 51(5): 655–91.

22 Black, B. A., Neely, R. R. and Manga, M. 2015. Campanian Ignimbrite Volcanism, Climate, and the Final Decline of the Neanderthals. *Geology* 43(5): 411–14.

23 Underdown, S. 2008. A Potential Role for Transmissible Spongiform Encephalopathies in Neanderthal Extinction. *Medical Hypotheses* 71(1): 4–7.

24 Whitfield, J. T. et al. 2008. Mortuary Rites of the South Fore and Kuru. *Philosophical Transactions of the Royal Society B* 363: 3721–4.

25 Alpers, M. P. 2007. A History of Kuru. *Papua and New Guinea Medical Journal* 50(1–2): 10–19.

26 Riel-Salvatore, J. 2008. Mad Neanderthal Disease? Some Comments on 'A Potential Role for Transmissible Spongiform Encephalopathies in Neanderthal Extinction'. *Medical Hypotheses* 71(3): 473–4.

27 Valet, J.-P. and Valladas, H. 2010. The Laschamp-Mono Lake Geomagnetic Events and the Extinction of Neanderthal: A Causal Link or a Coincidence? *Quaternary Science Reviews* 29(27–8): 3887–93.

28 Mellars, P. and French, J. C. 2011. Tenfold Population Increase in Western Europe at the Neandertal-to-Modern Human Transition. *Science* 333: 623–7.

29 Dogandžić, T. and McPherron, S. P. 2013. Demography and the Demise of Neandertals: A Comment on 'Tenfold Population Increase in Western Europe at the Neandertal-to-Modern Human Transition'. *Journal of Human Evolution* 64: 311–13.

30 Li, H. and Durbin, R. 2011. Inference of Human Population History from Individual Whole-Genome Sequences. *Nature* 475: 493–6.

31 Prüfer, K. et al. 2014. The Complete Genome Sequence of a Neanderthal from the Altai Mountains. *Nature* 505(7481): 43–9.

32 Meyer, M. et al. 2012. A High-Coverage Genome Sequence from an Archaic Denisovan Individual. *Science* 338: 222–6; Mafessoni, F. et al. 2020. A High-Coverage Neandertal Genome from Chagyrskaya Cave. *bioRxiv* 2020.03.12.988956. https://doi.org/10.1101/2020.03.12.988956.

33 Estimating ancient population size is extremely difficult. The most detailed analysis is Bocquet-Appel, J-P. and Degioanni, A. 2013. Neanderthal Demographic Estimates. *Current Anthropology* 54(8): S202–13, whose authors show ranges in Neanderthal population sizes based on nine different scenarios from between 5,000 and 70,000 individuals. More recent work based on nuclear genome data suggests that the lower end of the range is the most likely. Prüfer et al. (2014), for example, have suggested a figure of 1,000–5,000 individuals based on the Altai Neanderthal homozygosity and using the pairwise sequential Markovian coalescent (PSMC) model of Li and Durbin (2011). This method appears to predict known population histories from several populations rather well. Rogers, A. R. et al. 2017. Early History of Neanderthals and Denisovans.

Proceedings of the National Academy of Sciences 114(37): 9859–63, details a slightly different approach based on different gene trees and split times between populations. Its authors estimated a population of 15,000 Neanderthals, but this was criticized in Mafessoni, F. and Prüfer, K. 2017. Better Support for a Small Effective Population Size of Neandertals and a Long Shared History of Neandertals and Denisovans. *Proceedings of the National Academy of Sciences* 114(48) E10256–E10257, whose authors suggested that the method was at odds with the evidence from the low-coverage genomes previously published. Taken together, the current evidence suggests a population of between 1,000 and 5,000 is the most likely, but with the caveat that more high-coverage genetic data might change the estimate somewhat. See also Prüfer, K. et al. 2017. A High-Coverage Neandertal Genome from Vindija Cave in Croatia. *Science* 358: 655–8.

34 Prüfer et al. 2017.

35 Fu, Q. et al. 2014. The Genome Sequence of a 45,000-Year-Old Modern Human from Western Siberia. *Nature* 514: 445–9.

36 Sikora, M. et al. 2017. Ancient Genomes Show Social and Reproductive Behavior of Early Upper Paleolithic Foragers. *Science* 358: 659–62.

37 The way this is determined is based on the fact that in small populations it turns out that natural selection is less effective at removing mutations that are deleterious. One clue to measuring this is by looking at non-synonymous substitutions in the genome. These are mutations that introduce a single mutation in a nucleotide that changes the expression of a protein, which is obviously a serious effect. If one obtains the ratio of non-synonymous to synonymous substitutions then one can generate an estimate of how effective natural selection has been at removing these mutations and thereby obtain an indirect assessment of the past size of a population, since larger populations are affected by fewer non-synonymous substitutions (see Meyer et al. 2012).

38 Pearce, E. 2018. Neanderthals and *Homo sapiens*: Cognitively Different Kinds of Human? In L. D. Di Paolo, F. Di Vincenzo and F. De Petrillo (eds.), *Evolution of Primate Social Cognition*, Cham: Springer, pp. 181–96.

39 Vaesen, K. et al. 2019. Inbreeding, Allee Effects and Stochasticity Might be Sufficient to Account for Neanderthal Extinction. *PLoS ONE* 14(11): e0225117.

40 Kolodny, O. and Feldman, M. W. 2017. A Parsimonious Neutral Model Suggests Neanderthal Replacement was Determined by Migration and Random Species Drift. *Nature Communications* 8: 1040.

Chapter 16: Our Genetic Legacy

1 Simonti, C. N. et al. 2016. The Phenotypic Legacy of Admixture Between Modern Humans and Neandertals. *Science* 351: 737–41.

2 The Biobank has an online database at http://geneatlas.roslin.ed.ac.uk/ that allows visitors to search for specific phenotypes and explore the genome-wide association results for over 30 million genetic variants. Canela-Xandri, O., Rawlik, K. and Tenesa, A. 2018. An Atlas of Genetic Associations in UK Biobank. *Nature Genetics* 50: 1593–9.

3 Dannemann, M. and Kelso, J. 2017. The Contribution of Neanderthals to Phenotypic Variation in Modern Humans. *American Journal of Human Genetics* 101(4): 578–89.

4 Lalueza-Fox, C. et al. 2007. A Melanocortin 1 Receptor Allele Suggests Varying Pigmentation Among Neanderthals. *Science* 318: 1453–5.

5 Robles, C. et al. 2020. The Impact of Neanderthal Admixture on the Genetic Architecture of Complex Traits. Manuscript in review/submission.

6 Pearce, E. 2013. The Effects of Latitude on Hominin Social Network Maintenance. Unpublished DPhil dissertation, Oxford University.

7 Sankararaman, S. et al. 2014. The Genomic Landscape of Neanderthal Ancestry in Present-Day Humans. *Nature* 507: 354–7.

8 The delivery mechanism for Type II diabetes is a protein that moves certain lipids, or fatty acids, into the liver. In a study published in 2014, the Slim Initiative in Genomic Medicine for the Americas, or SIGMA, found through testing over 8,200 Mexican and other Latin American individuals that two protein coding genes (*SLC16A11* and *SLC16A13*) were significantly correlated with Type 2 diabetes (Latin American people have around the same amount of Neanderthal DNA as do people in wider Eurasia). Varying levels of the *SLC16A11* protein influence the presence of a type of fat implicated in the disease. SIGMA

Type 2 Diabetes Consortium, Williams, A. L. et al. 2014. Sequence Variants in *SLC16A11* are a Common Risk Factor for Type 2 Diabetes in Mexico. *Nature* 506(7486): 97–101.

9 Fu, Q. et al. 2016. The Genetic History of Ice Age Europe. *Nature* 534: 200–205.

10 Petr, M. et al. 2019. Limits of Long-Term Selection Against Neandertal Introgression. *Proceedings of the National Academy of Sciences* 116(5): 1639–44.

11 Gunz, P. et al. 2019. Neandertal Introgression Sheds Light on Modern Human Endocranial Globularity. *Current Biology* 29(1): 120–27.

12 Mallick, S. et al. 2016. The Simons Genome Diversity Project: 300 Genomes from 142 Diverse Populations. *Nature* 538: 201–6.

13 Chen, L. et al. 2020. Identifying and Interpreting Apparent Neanderthal Ancestry in African Individuals. *Cell* 180(4): 677–87. The proportions determined by Chen et al. are consistent with the low proportions of Neanderthal DNA that were documented previously by others, for example in the 1000 Genomes Project and in Prüfer, K. et al. 2014. The Complete Genome Sequence of a Neanderthal from the Altai Mountains. *Nature* 505(7481): 43–9.

14 Zammit, N. W. et al. 2019. Denisovan, Modern Human and Mouse *TNFAIP3* Alleles Tune A20 Phosphorylation and Immunity. *Nature Immunology* 20: 1299–1310.

15 Almarri, M. et al. 2020. Population Structure, Stratification, and Introgression of Human Structural Variation. *Cell* 182: 189–99. https://doi.org/10.1016/j.cell.2020.05.024.

16 Wu, T. et al. 2005. Hemoglobin Levels in Qinghai-Tibet: Different Effects of Gender for Tibetans vs. Han. *Journal of Applied Physiology* 98(2): 598–604.

17 Niermeyer, S. et al. 1995. Arterial Oxygen Saturation in Tibetan and Han Infants Born in Lhasa, Tibet. *New England Journal of Medicine* 333: 1248–52.

18 Moore, L. G. 2001. Human Genetic Adaptation to High Altitude. *High Altitude Medicine & Biology* 2(2): 257–79.

19 Like many proteins, the acronym reveals a complex name: Endothelial PAS domain-containing protein 1.

20 Huerta-Sánchez, E. et al. 2014. Altitude Adaptation in Tibetans Caused by Introgression of Denisovan-Like DNA. *Nature* 512: 194–7.

21 See interview with Emilia Huerta-Sanchez, 29 May 2019, at https://insitome.libsyn.com/natural-selection-and-deep-learning.

22 Yi, X. et al. 2010. Sequencing of 50 Human Exomes Reveals Adaptation to High Altitude. *Science* 329: 75–8.

23 Miao, B., Wang, Z. and Li, Y. 2017. Genomic Analysis Reveals Hypoxia Adaptation in the Tibetan Mastiff by Introgression of the Gray Wolf from the Tibetan Plateau, *Molecular Biology and Evolution* 34(3): 734–43.

24 Chen, N. et al. 2018. Whole-Genome Resequencing Reveals World-Wide Ancestry and Adaptive Introgression Events of Domesticated Cattle in East Asia. *Nature Communications* 9: 2337: https://doi.org/10.1038/s41467-018-04737-0; Wu, D.-D. et al. 2018. Pervasive Introgression Facilitated Domestication and Adaptation in the *Bos* Species Complex. *Nature Ecology and Evolution* 2: 1139–45.

25 Zhang, X. L. et al. 2018. The Earliest Human Occupation of the High-Altitude Tibetan Plateau 40 Thousand to 30 Thousand Years Ago. *Science* 362: 1049–51.

26 Racimo, F. et al. 2017. Archaic Adaptive Introgression in *TBX15/WARS2*. *Molecular Biology and Evolution* 34(3): 509–24.

Chapter 17: *The World Before Us*

1 Green, R. E. et al. 2010. A Draft Sequence of the Neandertal Genome. *Science* 328: 710–22.

2 Grün, R. et al. 2020. Dating the Skull from Broken Hill, Zambia, and Its Position in Human Evolution. *Nature* 580: 372–5. https://doi.org/10.1038/s41586-020-2165-4.

3 Rogers, J. et al. 2019. The Comparative Genomics and Complex Population History of *Papio* Baboons. *Science Advances* 5: eaau6947.

4 Ackermann, R. R. et al. 2019. Hybridization in Human Evolution: Insights from Other Organisms. *Evolutionary Anthropology* 28(4): 189–209.

5 Ibid.

6 Hajdinjak, M. et al. 2018. Reconstructing the Genetic History of Late Neanderthals. *Nature* 555: 652–6.

7 Kaifu, Y. et al. 2015. Unique Dental Morphology of *Homo floresiensis* and Its Evolutionary Implications. *PLoS ONE* https://doi.org/10.1371/journal.pone.0141614.

8 Martinón-Torres, M. et al. 2017. *Homo sapiens* in the Eastern Asian Late Pleistocene. *Current Anthropology* 58(S17): S434–S448.

9 Kolobova, K. A. et al. 2020. Archaeological Evidence for Two Separate Dispersals of Neanderthals into Southern Siberia. *Proceedings of the National Academy of Sciences* 117(6): 2879–85; Meyer, M. et al. 2012. A High-Coverage Genome Sequence from an Archaic Denisovan Individual. *Science* 338: 222–6.

10 Peter, B. M. 2020. 100,000 Years of Gene Flow Between Neandertals and Denisovans in the Altai Mountains. *bioRxiv* 2020.03.13.990523. https://doi.org/10.1101/2020.03.13.990523.

11 Allen, R. et al. 2020. A Mitochondrial Genetic Divergence Proxy Predicts the Reproductive Compatibility of Mammalian Hybrids. *Proceedings of the Royal Society B* 287: 2020.0690.

12 McBrearty, S. and Brooks, A. S. 2000. The Revolution That Wasn't: A New Interpretation of the Origin of Modern Human Behavior. *Journal of Human Evolution* 39: 453–563.

13 Greenbaum, G. et al. 2019. Was Inter-Population Connectivity of Neanderthals and Modern Humans the Driver of the Upper Paleolithic Transition Rather Than Its Product? *Quaternary Science Reviews* 217: 316–29.

14 Vernot, B. and Akey, J. M. 2014. Resurrecting Surviving Neandertal Lineages from Modern Human Genomes. *Science* 343: 1017–21.

Index

Page references in *italics* indicate images.